Perfluoroalkyl and Polyfluoroalkyl Substances

Perfluoroalkyl and Polyfluoroalkyl Substances: Environmental Fate, Health Impacts, and Remediation Strategies provides useful information on recent advancements in the use of perfluoroalkyl and polyfluoroalkyl substances, as well as their sources, environmental fate and conveyance, potential health impacts, and remediation strategies by harnessing the latest research findings and key scientific literature. The book draws its strength from a multidisciplinary approach with an emphasis on the enhancement of the understanding of perfluoroalkyl and polyfluoroalkyl substances and informs future research and management efforts.

- The book focuses on how perfluoroalkyl and polyfluoroalkyl substances can be used in the environmental and biomedical sectors to create sustainability.
- The book covers practical case studies, implementation guides, and the latest advancements and future trends in perfluoroalkyl and polyfluoroalkyl substances for environmental and biomedical sustainability.
- The book is unique in its comprehensive coverage of sustainability and its applications in sustainable environmental and biomedical science, as highlighted by the title.

Green Chemical Innovations and Sustainability – Series Editor, Professor Anja Mudring, Aarhus University, Denmark

This series is dedicated to bringing together academic researchers, industry professionals, teachers, law and policy makers, who constitute the growing green chemistry community. We welcome volumes on recent innovations, as well as accompanying educational resources from those working in related disciplines across chemistry, chemical engineering, and environmental science.

All contributions should clearly demonstrate impact to advancing the field and where possible reference the 12 Principles of Green Chemistry and the Sustainable Development Goals agreed by the United Nations in 2015. Volumes could cover:

- Green synthesis and catalysis
- Green manufacturing and engineering
- Green product design
- Green chemistry education
- Critical raw materials and sustainable resources
- Energy harvest, storage, and savings
- Circular economy
- Environmental chemistry and green metrics
- Ethics, legislation, and economics

Recent Advances in Activated Carbon: Synthesis, Properties and Applications
Ho Soon Min, Heri Septya Kusuma, and Yogesh Chandra Sharma

Perfluoroalkyl and Polyfluoroalkyl Substances

Environmental Fate, Health Impacts, and Remediation Strategies

Mohamed Aly Hassaan, Kingsley
Eghonghon Ukhurebor, Uyiosa Osagie Aigbe,
Robert Birundu Onyancha, Amr Gamal Dardeer,
Marwa Ramadan ElKatory, Kenneth Atoe, and
Ahmed El Nemr

CRC Press
Taylor & Francis Group
Boca Raton London New York

CRC Press is an imprint of the
Taylor & Francis Group, an **informa** business

First edition published 2026
by CRC Press
2385 NW Executive Center Drive, Suite 320, Boca Raton FL 33431

and by CRC Press
4 Park Square, Milton Park, Abingdon, Oxon, OX14 4RN

CRC Press is an imprint of Taylor & Francis Group, LLC

ISBN: 9781041038191 (hbk)
ISBN: 9781041038207 (pbk)
ISBN: 9781003625537 (ebk)

DOI: 10.1201/9781003625537

Typeset in Times
by Deanta Global Publishing Services, Chennai, India

Contents

About the Authors

Mohamed Aly Hassaan is an Associate Professor at the National Institute of Oceanography and Fisheries (NIOF) in Alexandria, Egypt. Dr. Hassaan holds a Master's degree (2011) and a PhD (2017) in Chemical Oceanography and Marine Pollution. He has lectured on marine science and environmental topics at Port Said University. In 2018, he completed a post-doctoral fellowship at Bari University, Italy. Dr. Hassaan has participated in multiple oceanographic expeditions, including those hosted by NIOF and Italy's National Research Council (CNR). He has completed specialized training courses at IHE Delft Institute for Water Education in the Netherlands (2015 and 2016) and was a visiting fellow at CNR in 2014 and 2017. Dr. Hassaan's prolific academic output includes co-authoring several publications, accumulating over 1,600 citations on Scopus. His research interests encompass advanced oxidation techniques (ozonation, sonication, ultraviolet), wastewater treatment, nanotechnology, biogas and biofuel energy production, monitoring of inorganic and organic pollutants, and the classification of Mediterranean minerals using PSA, FTIR, and XRD techniques. Dr. Hassaan is committed to pioneering innovative green technologies for pollutant treatment and sustainable energy production.

Kingsley Eghonghon Ukhurebor is currently a Senior Lecturer/Researcher (due to an Associate Professor) and the current acting Director, Centre for Open and Distance Learning and the immediate past acting Director of Sports and a former acting Head of the Department of Physics at Edo State University, Iyamho, Nigeria. He is also a Research Fellow at the West African Science Service Centre on Climate Change and Adapted Land Use (WASCAL), Competence Centre, Ouagadougou, Burkina Faso, which is a climate institute sponsored by the Federal Ministry of Education and Research, Germany. He was awarded his PhD in Physics Electronics by the University of Benin, Benin City, Nigeria; his MSc in Physics Electronics by the University of Benin, Benin City, Nigeria; his PGD in Education by Usmanu Danfodiyo University, Sokoto, Nigeria; and his BSc in Applied Physics by Ambrose Alli University, Ekpoma, Nigeria. He is a member of several learned academic organizations, such as the Nigerian Young Academy (NYA), the Nigerian Institute of Physics (NIP), the Materials Science and Technology Society of Nigeria (MSN), and the Teacher Registration Council of Nigeria (TRCN). His research interests include applied physics, climate physics, environmental physics, telecommunications physics, and materials science (nanotechnology). He serves as an editor and reviewer for several reputable journals and publishers, including Springer Nature, Elsevier, the Royal Society of Chemistry (RSC), the Institute of Physics (IOP), Taylor & Francis, John Wiley & Sons, the IEEE, MDPI, Frontiers Media SA, Hindawi, and he acts as a supervisor and examiner for undergraduate and postgraduate students. He has authored or co-authored several publications with these reputable journals and publishers. He is currently ranked among the top 50 authors in Nigeria by Scopus scholarly output and is listed among the top 2% of scientists in the world by Stanford

University, USA, and Elsevier. His hobbies include reading and playing/watching football. He is a Christian and is happily married with a family.

Uyiosa Osagie Aigbe is a former Research Fellow with the Department of Mathematics and Physics, Faculty of Applied Science, Cape Peninsula University of Technology, Cape Town, South Africa, and is currently a Research Associate with the Center for Space Research, North-West University, Potchefstroom, South Africa, the National Institute of Theoretical and Computational Science, Johannesburg, South Africa. He obtained his PhD degree in physics from the prestigious University of South Africa, Pretoria, South Africa. He is currently a member of several learned academic organizations. His research interests are in applied physics, nanotechnology, fluid dynamics, water purification processes, image processing, environmental physics, machine learning, statistical analysis, and materials science. He has also served as a reviewer and editor for numerous highly regarded publishers, such as Elsevier, Springer Nature, the Royal Society of Chemistry (RSC), the Institute of Physics (IOP), Taylor & Francis, Frontiers Media SA, MDPI, and Hindawi, and has authored or co-authored several research publications. His hobbies are reading, swimming, and playing/watching football. He is a Christian and is happily married with a family.

Robert Birundu Onyancha is a Senior Lecturer of Physics at the Technical University of Kenya, a research Fellow at the College of Graduate Studies, School of Interdisciplinary Research and Graduate Studies, University of South Africa, and a senior lecturer of physics on sabbatical at the Department of Physics, School of Pure and Applied Science, Kisii University, Kenya. He holds a Ph.D. (Physics) and an MSc (Physics) from the University of South Africa. His research interests are in material science, wastewater treatment technologies, superconductivity, and magnetism. He has authored and co-authored numerous research papers, book chapters, and published books. Furthermore, he serves as an editor and reviewer of highly accredited journals and supervises master's and PhD students in the above-listed research interest fields.

Amr Gamal Dardeer is a distinguished chemistry specialist with a focus on environmental monitoring and water quality. He previously served as the manager of the Administration of Environmental Monitoring at the Alexandria Health Directorate, part of Egypt's Ministry of Health and Population. In this role, he oversaw public health protection initiatives through effective environmental management. In addition to his governmental duties, Dr. Gamal manages the inorganic chemistry laboratory at the Environmental Measurements Company (EMCO), where he leads efforts to assess water quality and soil health. His expertise is essential for ensuring compliance with environmental standards and promoting sustainable practices. Dr. Gamal earned his Master's degree in 2012 and completed his PhD in 2019, enhancing his research and knowledge in chemistry. His contributions have earned him respect in both academic and professional circles, where he continues to advance environmental health in Egypt.

Marwa Ramadan Elkatory is a chemist at the City of Scientific Research and Technology Application (SRTA city) in Alexandria, Egypt, where she has been a dedicated research member since 2009. Dr. Elkatory holds a Master's degree (2012) and a PhD (2017) in Microbiology and Organic Chemistry. Her prolific academic output includes co-authoring three book chapters and 31 research articles, accumulating over 686 citations on Scopus. Her research interests encompass genetic engineering, bioplastic, biopolymers, additives engine, cold flow improvers for petroleum crude oil wastewater treatment, nanotechnology, biodiesel, biogas, biofuel energy, and green hydrogen production. Dr. Elkatory is committed to pioneering innovative green technologies for industrial problems and sustainable energy production.

Kenneth Atoe holds a BSc and MSc in Biochemistry from Ambrose Alli University Ekpoma and University of Benin, Benin City, Nigeria, respectively. He also obtained his Bachelors in Medicine and Surgery and a PhD in Phytomedicine from University of Benin, Benin City. Dr. Atoe is the immediate past acting Provost of the College of Medical Sciences, Edo State University Uzairue,Iyamho, Nigeria. He loves to research and write, with review and editorship experience with numerous international publications. He has been featured as an invited speaker for reputable associations and bodies. Dr. Atoe volunteers as a reviewer for many double-blind peer review journals, hosted by reputable publishers. He is well published, with scores of research articles to his credit in peer-reviewed journals and platforms. Dr. Atoe loves listening to music, travelling to new places, and meeting new people. He is married to Mrs. Wisdom Atoe and is blessed with three children.

Ahmed El Nemr is a full Professor of Chemistry at the National Institute of Oceanography and Fisheries (NIOF). One of Dr. El Nemr's interests is exploring new approaches for synthetic methodologies in carbohydrate chemistry and the synthesis of natural compounds, the isolation of some natural compounds from marine algae and has tested for their use as antivirals and anti-corrosion, the synthesis of cellulose acetate using new methods. Dr. El Nemr has also studied the preparation of membranes for desalination and wastewater treatment, quantum chemical calculations for deducing the reaction mechanism of new reactions and to examine the "quantum structure–activity relationship", the development of novel activated carbon from agricultural waste, the removal of textile dyes, organic pollutants and heavy metals from water using macroalgae, activated carbon, biochar, hydrogel and agricultural wastes, synthesizing metal nanocomposites and photodegradation studies for wastewater treatment. Biogas, biomethane, and hydrogen generation using nanomaterials and visible light. The distribution of petroleum hydrocarbons, pesticides, polycyclic aromatic hydrocarbons, polychlorinated biphenyls, and heavy metals in the sea have been investigated, and these have been linked to pollution and environmental aspects. Stanford University has ranked Dr. El Nemr within 2% of scientists worldwide for six consecutive years. He has supervised and refereed several Master and PhD theses. Dr. El Nemr is a permanent referee in dozens of international journals and is a co-editor and reviewer for several reputable publishers. He has published 10 international books and more than 240 international research papers in various fields.

1 An Overview of Perfluoroalkyl and Polyfluoroalkyl Substances

1.1 INTRODUCTION

At the moment, emerging pollutants that contain significant levels of both natural and man-made sources have become a major concern in a number of spectrums of the environment, especially the aquatic ecosystems (Aidonojie et al., 2023; Nwankwo et al., 2023; Ukhurebor et al., 2023a,b, 2022; Kerry et al., 2022). The main cause of aquatic contaminants is large-scale commercial and industrial advancements (Ukhurebor et al., 2024; Ukhurebor & Aigbe, 2024; Anani et al., 2023; Aigbe et al., 2023; Singh et al., 2022). The environment and public health sectors are greatly concerned about perfluorinated substances or perfluoroalkyl and polyfluoroalkyl compounds or substances (PFAS), often known as forever chemicals, which constitute some of the primary sources of pollution in the environments, especially water-based ecosystems (Gewurtz et al., 2024; Cousins et al., 2022; Fenton et al., 2021; Elmoznino et al., 2018).

PFAS are a broad class of about 4,000 partly or completely fluorinated linear, branching, or cyclic chemicals (Gewurtz et al., 2024). As adapted from the Organisation for Economic Cooperation and Development (OECD), the broad classes of the main compounds of PFAS using a nomenclature (classification system) for PFAS that are generally recognized are shown in Figure 1.1 (OECD, 2015). They are a class of man-made compounds and since their discovery in the late 1940s, they have grown to be a crucial component of industrial and consumer goods all over the world. PFAS have been utilized in numerous household products and industrial processes since the 1950s (Gewurtz et al., 2024; Obodo, 2024; Sunderland et al., 2019). They have qualities that make them resistant to water, grease, and heat. PFAS come in many different types. According to a report from the US Environmental Protection Agency (EPA), perfluorooctanoic acid (PFOA) and perfluorooctanoic sulfonic acid (PFOS) are the two most prevalent varieties (EPA, 2014; EPA, 2017). Due to their historical applications, they can still be found in the environment and in certain fire-fighting foams. Furthermore, additional PFAS are frequently used in products in place of PFOA and PFOS.

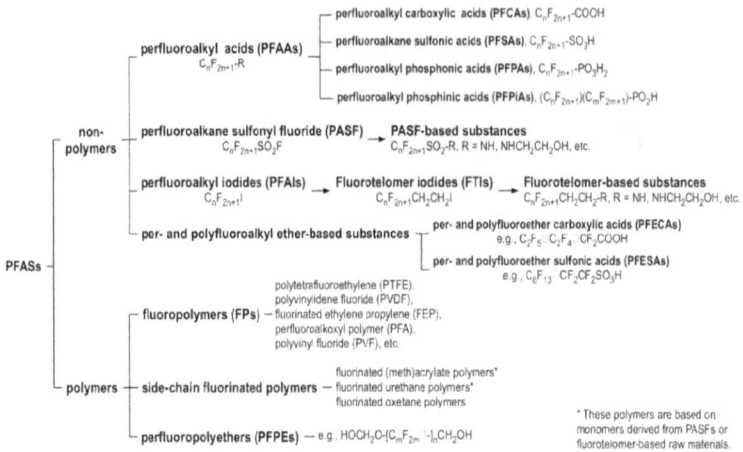

FIGURE 1.1 The broad classes of main chemicals or compounds of PFAS using a nomenclature (classification system) for PFAS that are generally recognized. Adapted and reproduced from the Organisation for Economic Co-operation and Development (OECD) with permission from OECD open access article distributed under the terms and conditions of the Creative Commons Attribution (CC BY) license.

They are an active component of adhesives, foams used to fight fires, cosmetics, and paper goods, and are employed in the textile sector as water- and stain-repellents. PFAS are utilized in the production of agricultural items, coating additives, insecticides, lubricants, semiconductors, surfactants, and aqueous film-forming foams (AFFFs) for fire-fighting procedures (making up 1.00–5.00% w/w). The aviation sector also utilized PFAS as erosion inhibitors (Aminot et al., 2019; Oliaei et al., 2013; Zhang et al., 2021a, b; Sunderland et al., 2019). Owing to their hydrophobic and oleophobic characteristics, PFAS are also present in a wide range of home products, including food packaging and kitchenware. As reported by Glüge et al. (2020), the textile sector uses the greatest amount of PFAS and their precursors, followed by the packaging of papers as well as aftermarket consumer goods. Cleaning ingredients, water-resistant textiles (such as rain garments, umbrellas, and tents), non-stick kitchen equipment, grease-resistant paper products, items for personal care (such as shampoo and conditioner, floss for dental hygiene, cosmetics for the eyes, and nail polish), and stain-resistant coatings for flooring, furniture upholstery, and other textiles are all examples of frequently used goods that contain these PFAS. PFAS exposure primarily occurs from consuming tainted general population or individual well water, consuming seafood that contains a lot of PFAS, consuming food produced or grown close to facilities that utilize or manufacture PFAS, consuming food that has been packaged in PFAS-containing materials, ingesting polluted dust or soil, utilizing various consumer goods such as non-stick cooking equipment, ski wax-type substances, and cloth stain and water-repellent sprays (Glüge et al., 2020; Dawson et al., 2023; Redmon et al., 2025).

The majority of PFAS chemicals have not undergone enough evaluation, and it is unclear how they may negatively affect ecosystems or human health (Buck et al., 2011). As reported by the Environment and Climate Change Canada and Health Canada (ECCC and HC) as well as the Interstate Technology and Regulatory Council (ITRC), PFAS exposure may happen in a number of ways, but how bad it gets depends on how close the person is to the source, how much of the substance is involved, and how often the person is subjected to it (ECCC and HC, 2023a, b; ITRC, 2022). PFAS have been found in samples of human blood, in addition to samples of l terrestrial and and aquatic flora and wildlife. Some other studies, such as Gebbink and van Leeuwen (2020), Ahrens and Bundschuh (2014), and Quinete et al. (2009), have also ascertained these reports. To find out how PFAS harm us, scientists study both humans and animals. Human research has examined a potential connection between blood levels of PFAS and adverse health consequences. But only a few compounds have been examined in the majority of investigations. The adverse health consequences of PFAS are varied (ECCC and HC, 2023a,b; Cookson & Detwiler, 2022; Johnson, 2022; Gallen et al., 2022; Gottschall et al., 2017). According to research, higher concentrations of certain PFAS may raise cholesterol, impair the body's ability to respond to vaccinations, raise the possibility of thyroid disease, raise the risk of certain cancers, raise the risk of severe health conditions like high blood pressure or pre-eclampsia during pregnancy, and lower the delivery weight of a baby (though this weight loss is slight and may not have an impact on health). PFAS have been investigated in laboratory specimens in other studies. Animals and humans occasionally fail to respond to PFAS in the same manner.

PFAS are well known for their ability to persist in the environment and their connections to health issues. They are now being found in unexpected areas, including beer. According to Redmon et al. (2025), this phenomenon has not yet been examined in US retail beer. The researchers discovered a substantial link between PFAS concentrations in municipal water sources and levels in locally made beer; 95% of the beers they tested contained PFAS. These include two permanent compounds with recently set EPA limits in water for consumption: PFOA and PFOS. Of particular note, the study discovered that the largest concentrations and most varied mixture of forever chemicals, such as PFOS and PFOA, were identified in beers made close to the Cape Fear River Basin in North Carolina, an area known to be contaminated by PFAS. The researchers urge increased awareness among brewers, consumers, and regulators to prevent general PFAS exposure, as their research demonstrates how contamination with PFAS at one source can diffuse into other products. These findings also point to the potential necessity of brewing operations to adapt their water treatment procedures as water consumption PFAS rules change or as municipal water source processing is improved. However, there are methods for scientists to compare the outcomes in animals to those in humans. They make decisions on how to shield individuals from chemical harm based on the understanding they gain from this process. According to research on animals, PFAS are capable of having an impact on the functioning of the liver, of the immune system, and of the development in this aforementioned aspects (Glüge et al., 2020; Dawson et al., 2023; Redmon et al., 2025).

An Overview of Perfluoroalkyl and
Polyfluoroalkyl Substances

Sources and Production of
Perfluoroalkyl and
Polyfluoroalkyl Substances

Perfluoroalkyl and
Polyfluoroalkyl
Substances:
Environmental Fate,
Health Impacts, and
Remediation Strategies

Analytical Methods and
Monitoring for
Perfluoroalkyl and
Polyfluoroalkyl
Substances

Environmental Fate and
Transport of Perfluoroalkyl and
Polyfluoroalkyl Substances

Remediation Strategies
for Perfluoroalkyl and
Polyfluoroalkyl
Substances

Knowledge Gaps and Future
Research Directions for
Perfluoroalkyl and Polyfluoroalkyl
Substances

Regulations and Guidelines for
Perfluoroalkyl and
Polyfluoroalkyl Substances

Pathways of Human
Exposure to Perfluoroalkyl
and Polyfluoroalkyl
Substances and Health
Effects

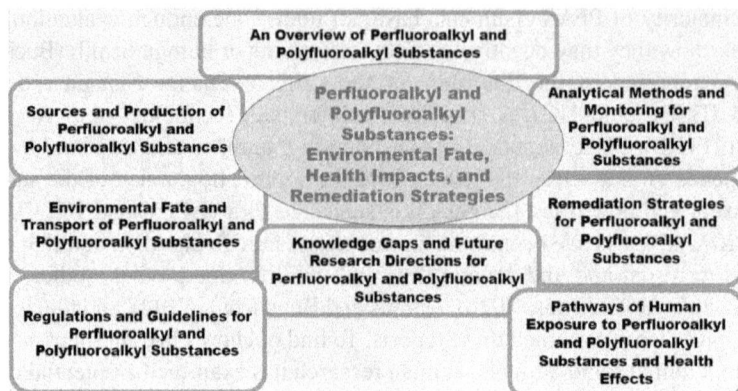

FIGURE 1.2 Illustrative summary of this book, *Perfluoroalkyl and Polyfluoroalkyl Substances: Environmental Fate, Health Impacts, and Remediation Strategies.*

Hence, a brief overview of the physical and chemical characteristics of PFAS, issues related to health and the environment, applications and possible environmental sources of PFAS, sampling and analytical techniques for PFAS, the fate and transport (pathways) of PFAS, and remediation strategies for PFAS are all highlighted in this introductory chapter (Chapter 1). Additionally, Chapter 2 attempts to deliberate the various sources and production of PFAS. Chapter 3 highlights the environmental fate and transport of PFAS, while Chapter 4 discusses the regulations and guidelines for PFAS. Chapter 5 deals with the analytical methods and monitoring for PFAS. The remediation strategies for PFAS are discussed in Chapter 6, followed by Chapter 7, which deals with the pathways of human exposure to PFAS and health effects, while the final chapter, Chapter 8, deals with the knowledge gaps and future research directions for PFAS. In summary, Figure 1.2 depicts the book's scheme, structure, and arrangement.

1.2   BRIEF PHYSICAL AND CHEMICAL CHARACTERISTICS OF PFAS

As reported by the Interstate Technology and Regulatory Council (ITRC), due to their polar and non-polar frameworks, PFAS are "amphiphilic", meaning they can bind to both water and oils (ITRC, 2018). They also have very high chemically and thermally stable states because of the strength of their carbon-fluorine interactions (Glüge et al., 2020; Dawson et al., 2023; Redmon et al., 2025). PFAAs are present in water-soluble anionic (i.e., deprotonated, negatively charged) states in the majority of groundwater as well as surface habitats. Under normal environmental circumstances, other PFAS groups may be cationic (which are recognized to be positively charged) or zwitterionic (which are recognized to have both a negative as well as positive charge). There are generally few physical characteristics of PFAS that have been reported, and what is known about PFAAs is mostly about their acid counterparts, which are not normally present in the environment (ITRC, 2018). The prediction of PFAAs' physiochemical characteristics, including partitioning coefficients and vapour pressure, is made more difficult by their surfactant characteristics. More details of some of the notable associated characteristics of PFOS and PFOA are provided in the report of the ITRC (2018). The ITRC information sheet on PFAS standard names and both chemical and physical characteristics has a description of the general features of PFAS (ITRC, 2018).

1.3 ISSUES RELATED TO THE HEALTH AND ENVIRONMENT, APPLICATIONS, AND POSSIBLE ENVIRONMENTAL SOURCES OF PFAS

A detailed study by Gebbink and van Leeuwen (2020) ascertained the existence of PFOA in samples of blood from locals residing close to a fluoro-chemical processing plant (FPP) in the Netherlands, with amounts getting up to 147.00 ng L^{-1}. In a similar vein, the North Carolina Department of Health and Human Services reported that the blood serum of persons residing close to an FPP was found to contain PFOA (of about 0.40–7.30 μg L^{-1}; median value = 1.75 μg L^{-1}) and PFOS (of about 1.40–34.60 μg L^{-1}; median concentration = 5.40 μg L^{-1}). Reduced spleen and thymus weights and cellularity, decreased generation of certain antibodies, lower survival following infections from influenza, immunosuppression, as well as modified production of cytokine are all possible outcomes of PFAS exposure (Grandjean et al., 2012). There are hardly any restrictions on the utilization of PFAS in consumer goods or industrial applications, notwithstanding the fact that there have been possibilities that they might be responsible for causing cancer as well as several other human diseases. Furthermore, because PFAS have extended persistence durations, bioaccumulation of these substances is simple. They are included in the list of new environmental pollutants (Bai & Son, 2021; Li et al., 2010).

Aquatic species are among the flora and animals that are threatened by PFAS. Because of the wide variations in each species' habitat, exposure, as well as the bioaccumulation of PFAS within the natural food web, it is challenging to specifically estimate environmental concerns (Ahrens and Bundschuh, 2014). Furthermore, the internal organs of different animals have varying levels of different PFAS accumulation. While PFOA may accumulate in muscle tissues above the kidney, PFOS can accumulate in greater amounts in the liver than in muscle tissues (liver > muscle > kidney). Aquatic species differ in their bioaccumulation rates as well; mussels accumulate PFOA in greater amounts than fish (Quinete et al., 2009). Mean PFOS concentrations of approximately 2000 ng g^{-1} have been found in different species of aquatic animals, including polar bears as well as several other species of dolphins, porpoises, seals, and whales (Houde et al., 2011). PFAS's long-term environmental persistence may have effects on the food chain at several different levels (Goodrow et al., 2020; Ghisi et al., 2019; Corsini et al., 2014).

Given their high degree of environmental consistency and movement ability, widespread detection, and possible toxicological consequences on both humans and the environment, PFAS are a source of problems for the environment (Glüge et al.,

2020; Dawson et al., 2023; Redmon et al., 2025). PFAS do not degrade in the environment and are extremely stable, especially PFAAs. These substances are, therefore, present in all parts of the world's environment. Trace quantities of PFAS have been identified in remote areas like the Arctic, far from probable point sources (Young et al., 2007). Certain long-chain PFAS bioaccumulate and biomagnify in wildlife, according to other research (Conder et al., 2008). As a result, humans that consume higher trophic organisms, such as aquatic creatures and birds, may be especially vulnerable to any harmful consequences for health that PFAS may produce (Sinclair et al., 2006). Based on human fish consumption, the Dutch National Institute for Public Health and the Environment determined that the maximum allowable concentration of PFOS in water from fresh sources is 0.65 nanograms per litre (ng/L) (EPA, 2014). As reported by the Minnesota Department of Health (MDH), there have been recent cautions regarding the consumption of aquatic organisms in specific US areas linked to PFAS pollution (MDH, 2018).

According to a report from the Agency for Toxic Substances and Disease Registry (ATSDR), the toxicological consequences of PFAS on people are still being studied, much like other aspects of PFAS study. In humans, the half-lives of PFOA and PFOS are 2.10 to 8.50 years and 3.10 to 7.40 years, respectively (ATSDR, 2018). PFAS usually build up in the circulation, liver, and proteins (EPA, 2017). Studies on the toxicity and epidemiology of PFOA, PFOS, and other PFAAs suggest that they may be linked to a number of illnesses, such as reduced fertility, elevated cholesterol, immune system suppression, and several types of cancer (ATSDR, 2018). Although PFOA and PFOS have been identified as carcinogens, the US EPA has not yet determined how carcinogenic they are (EPA, 2017). According to the International Agency for Research on Cancer (IARC), PFOA is a Group 2B carcinogen, meaning it may cause cancer in people (IARC, 2016; Benbrahim-Tallaa et al., 2014). For both PFOA and PFOS, the US EPA released draft oral reference levels of 20.00 ng/kg-day (based on non-cancer hazards) (EPA, 2017). The exposure pathways that are dangerous for the health of humans are the consumption of drinking water and fish, exposure to water through the skin, and (unintentional) ingestion or contact with soil that is polluted.

According to a recent report by Quinete and Ogunbiyi (2024), the water consumed is not the only source of PFAS; another source is the "forever chemicals", which reportedly have been causing health issues globally. These harmful substances can wind up in streams when they leak out of malfunctioning wastewater treatment plants, landfills, and fields of agriculture that eventually empty into aquatic environments that are home to species such as fish, dolphins, manatees, sharks, and other aquatic animals. The proliferation of PFAS has an impact on both human health and the environment since these chemicals can find their way into the food chain and build up in aquatic organisms and creatures, particularly aquatic animals such as fishes that humans consume (Ehsan et al., 2024; Quinete and Ogunbiyi, 2024). According to Ehsan et al. (2024), all stages of the food chain have been shown to contain PFAS in organic substances, water, sediment, and aquatic organisms. To fully comprehend the extent of PFAS contamination, however, a significant number of studies need to be done. The most predominant sources or mediums by which

PFAS get into the environment are feasibly grouped into two distinct categories: direct and indirect. Industries, metal coating processing, aqueous film-forming foam applications, paper and textile coating, etc. are some more categories into which the direct sources can be categorized. Leachate from landfills and waste from the treatment of wastewater facilities are examples of indirect sources (Dasu et al., 2022). According to the North Carolina Department of Health and Environmental Control (NCDEC), some of the predominant sources of PFAS in the environment are shown in Figure 1.3 (Quinete & Ogunbiyi, 2024).

In the past few decades, a number of review studies on the prevalence, exposure, and dispersion of PFAS have been published. For instance, Winchell et al. (2021), Schulz et al. (2020), Vo et al. (2020), Sunderland et al. (2019), Jian et al. (2017), Xiao (2017), Rahman et al. (2014), Ahrens and Bundschuh (2014), Ahrens (2011), Houde et al. (2011), as well as Rayne and Forest (2009) examined the incidence, fate, impact, exposure, and field analytical techniques of PFAS within the marine ecosystem (environment). Other researchers such as Wang et al. (2020a, b), Banzhaf et al. (2017), Wang et al. (2015), and Chen et al. (2009) have examined comparable PFAS elements at the regional level. Additionally, Zhang et al. (2021), Wanninayake (2021), Mahinroosta and Senevirathna (2020), Zhang et al. (2019a, b), Kucharzyk et al. (2017), Espana et al. (2015), Du et al. (2014), Rahman et al. (2014), and Parsons et al. (2008) have examined the confiscation of persistent organic pollutants (that is, PFAS) from soil and water, with particular attention to adsorption, bioremediation, nanotechnology, and innovative remediation techniques, as well as their comparative analysis. Each of these review publications addressed a distinct facet of PFAS, frequently concentrating on a particular area, environmental medium, or remediation

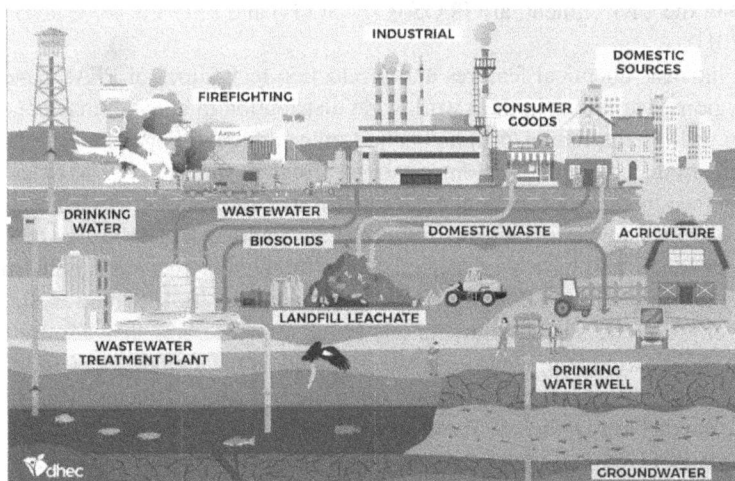

FIGURE 1.3 Some of the predominant sources of PFAS in the environment. Adapted and reproduced with permission from Quinete and Ogunbiyi (2024) with permission from the conversation open access article distributed under the terms and conditions of the Creative Commons Attribution (CC BY) license.

technique. At the continental level, none of these review publications offered a detailed analysis of PFAS. Thus, the goal of this book is to present a detailed analysis of the prevalence, distribution, regulation, and health hazards related to PFAS. In this work, the prevalence and spread of PFAS in the aquatic environment worldwide are examined through a comprehensive evaluation of studies released during the past ten years (Glüge et al., 2020; Dawson et al., 2023; Redmon et al., 2025). In addition, we assess the emergence of new PFAS and the developing international regulatory reaction to control and lessen the harmful risk PFAS as regards to human health. In the process, we offer vital details about the prevalence, distribution, and legal environment of PFAS worldwide. Because it does not concentrate on any one area or environmental media, this assessment is exceptional in its breadth of coverage.

PFAS are a broad group of compounds sharing an aliphatic carbon backbone with either totally or partly replaced hydrogen atoms by fluorine. Their durable, chemically inert carbon-fluorine (C–F) bonds and high polarity confer upon them special chemical properties, such as extraordinarily enormous chemical and thermal stability (Fernandez et al., 2016; Rahman et al., 2014). Because of these special characteristics, PFAS are extremely persistent, recalcitrant, and stable compounds, which makes it challenging to control their presence in a variety of environmental media, including soil, water, and air. For the sake of simplicity, this research employs a categorization scheme that describes PFAS as aliphatic compounds that include the $-C_nF_{2n+1}$ ($n \geq 1$) moiety. There are several nomenclature standards for PFAS groups (Buck et al., 2012). Furthermore, they are commonly divided into short-chain acids ($-CnF_{2n+1}$, $n \leq 6$) and long-chain acids (PFCAs, $C_nF_{2n+1}COOH$, $n \geq 7$), as well as their precursors, perfluoroalkyl sulfonic acids (PFSAs, $C_nF_{2n+1}SO_3H$, $n \geq 6$) (Yao et al., 2020). Among the most widely used PFAS, as well as the most often found and studied in the environment, are PFOS ($C_8F_{17}SO_3H$) and PFOA ($C_7F_{15}COOH$) (Buck et al., 2011).

The intrinsic chemical features and production techniques of PFAS have led to large regional variations in their worldwide distribution in aquatic (marine) ecosystems as well as the tissues of humans and other animals. For instance, compared to PFAS produced by fluorotelomerization, a region where PFAS were primarily produced using the conventional "electrochemical fluorination (ECF) technique" (which was abandoned in the United States in 2011 and results in bifurcated isomers as a by-product) will have a greater ratio of bifurcated isomers as an environmental noxious waste (Baabish et al., 2021; Sharifan et al., 2021; Schulz et al., 2020). Moreover, differences in structure among PFAS influence how they are transported through the environment. However, the bifurcated isomers tend to stay in the aqueous state, linear PFAS prefer to partition into the soil and sediments. This variation in how they behave in the environment may be used as an extra tool to determine the source of PFAS and how long they last (ITRC, 2020; Schulz et al., 2020). The environmental incidence and distribution of PFAS are determined by the length of their chain length as well as their functional moieties. While hydrophilic short-chain PFAS (of about $C \leq 8$) are usually set up on the surface of the waters, the hydrophobic long-chain PFAS (of about $C > 8$) tend to bioaccumulate in tissues of aquatic animals such as fish and the sediments (Bai & Son, 2021; Goodrow et al., 2020).

1.4 SAMPLING AND ANALYTICAL TECHNIQUES FOR PFAS

Sampling for PFAS necessitates cautionary measures to prevent contamination by others from additional possible sources of PFAS because of the insufficient reporting levels required to compare to existing regulatory requirements and the existence of PFAS in several popular products for customers and sampling equipment. The majority of standard operating techniques and work plans caution against using fluoropolymer-based (such as Teflon) materials and taking extra care with sample containers, sampler clothes, as well as the handling of some commonplace objects. Industrial laboratories use the EPA-approved technique 537.10, which combines liquid chromatography and solid-phase extraction with tandem mass spectrometry, to assess PFAS in the consumption of water specimens (EPA, 2019a). A modified form of 537.10 is required to quantify around 24 distinct PFAS chemicals in samples taken from mediums other than water for human consumption. In the summer of 2019, the EPA published technique for some of these other media, for comments from the public (EPA, 2019b). Up to 40 chemicals might be added to the target analyte list by certain industrial laboratories. The Total Oxidizable Precursor (TOP) Assay is another analytical technique that certain industrial laboratories provide. It measures the mass of PFAS in a specimen in large quantities, including oxidizable precursors (Houtz et al., 2013; Houtz & Sedlak, 2013). Other methods for measuring the overall concentration of organic fluorine in water specimens comprise absorbable organic fluorine (AOF) and particle-induced gamma-ray emission (PIGE) (Birnbaum & Grandjean, 2015).

1.5 FATE AND TRANSPORT (PATHWAYS) OF PFAS

The fate and transport (pathways) of PFAS can be influenced by the sorption and biotransformation processes, as well as the existence of co-contaminants. The long-chain composites have a more dominant affinity when compared to the short-chain composites, and PFSA has a more durable affinity when compared to the PFCAs for a particular chain length (Higgins & Luthy, 2006). It has been noted that PFAS have varied degrees of affinity for "solid-phase organic carbon", depending on the chain length and arrangement. Under some circumstances, the interactions with mineral stages (specifically, ferric oxide materials) may potentially be significant (Ferrey et al., 2012; Johnson et al., 2007). To effectively anticipate PFAS mobility, empirical site-specific sorption values are currently recommended (Ferrey et al., 2012). In the environment, PFAS can be challenging to break down. PFOA and PFOS are examples of "terminal", resistant PFAA forms that can be produced by biotic or abiotic degradation of polyfluorinated forms, as well as other intermediate forms (Harding-Marjanovic et al., 2015; Tseng et al., 2014; Ellis et al., 2004). Due to this, these biodegradable PFAS are occasionally called "precursors" of PFAA. Precursor decomposition can be accelerated by co-contaminant remediation, especially when oxidation-based methods are used. PFAS migration may be slowed down by interactions with non-aqueous phase liquids.

1.6   REMEDIATION STRATEGIES FOR PFAS

Remediation strategies for PFAS in soil and groundwater are expensive and challenging simply because of the thermal and chemical resistant properties of PFAS,
as well as the intricate nature of PFAS combinations of steps (Gewurtz et al., 2024;
Obodo, 2024). Effective corrective measures are currently being researched. Soil
treatment techniques include; direct on-site reusing and/or treatment, transitory on-
site preservation, and off-site discharge to a permitted landfill, incinerator, or soil
process or remediation facility (Glüge et al., 2020; Dawson et al., 2023; Redmon et
al., 2025). In situ management, ex situ management and/or recycling, reusing aquifer reinjection, release to the surface of the water, storm-water, or sewer, transitory
on-site storage, and off-site management which are potentially harmful for the management of waste or elimination plant are the available choices for management for
groundwater. Pump-and-treat using granular activated carbon (AC), accompanied
by off-site burning of the residual AC, is the most often used remediation technique.
Full-scale application of this technology has been going on for years (Appleman et
al., 2014). Nevertheless, in the presence of shorter chain chemicals, granular AC can
only absorb a small amount of PFAS. Tests for improving the capacity of sorption
have been performed on a variety of sorbent substances, including ion exchange resins, granular and granulated, clay, granulated AC, and other sorbent mixes (Du et al.,
2014). Reverse osmosis or high-pressure membrane remediation employing nanofiltration are further techniques for the ex situ detoxification of PFAS (Appleman et al.,
2014; Steinle-Darling & Reinhard, 2008). Presently, more investigations exploring
additional PFAS treatment techniques, such as enhanced oxidation techniques, biological approaches , in situ chemical reduction, and in situ barriers (sequestration),
are ongoing (Gewurtz et al., 2024).

1.7   CONCLUSION

In recent years, PFAS has been attracting a lot of public attention, and this interest
keeps intensifying. This is partly because research has shown that even relatively
small levels of PFAS exposure can have a wide range of adverse health and environmental consequences.

PFAS are found across several commercial and industrial commodities, have
been found in environmental media, can be hazardous to the health of humans and
the ecosystem, and could prove challenging to completely eliminate from the environment. They may bioaccumulate and biomagnify in the ecosystems, are extremely
transportable and stable in the natural environment, and are the focus of all levels of
government and governmental regulatory activities and legal actions. Health-based
water consumption recommendations have low concentrations (ng/L). Management,
as well as research personnel, are exploring locations, developing analytical methods, and developing and maintaining remediation strategies as regulatory requirements and knowledge of PFAS proliferate.

The focus of contemporary research is to increase the cost-effectiveness of technological innovations and demonstrate effective remediation solutions for PFAS.

Hence, this book will be a useful source of information on recent advancements in the use of PFAS, as well as their environmental fate and potential detrimental health effects on humans. The book draws its strength from a multidisciplinary approach with an emphasis on the enhancement of the understanding of PFAS and informs future research and management efforts.

REFERENCES

Agency for Toxic Substances and Disease Registry (ATSDR). (2018). *ToxGuide™ for perfluoroalkyls*. Available online: https://www.atsdr.cdc.gov/toxguides/toxguide-200.pdf (Accessed 20 May, 2025).

Ahrens, L. (2011). Polyfluoroalkyl compounds in the aquatic environment: A review of their occurrence and fate. *Journal of Environmental Monitoring, 13*(1), 20–31.

Ahrens, L., & Bundschuh, M. (2014). Fate and effects of poly- and perfluoroalkyl substances in the aquatic environment: A review. *Environmental Toxicology and Chemistry, 33*(9), 1921–1929.

Aidonojie, P. A., Ukhurebor, K. E., Oaihimire, I. E., Ngonso, B. F., Egielewa, P. E., Akinsehinde, B. O., Heri, S. K., & Darmokoesoemo, H. (2023). Bioenergy revamping and complimenting the global environmental legal framework on the reduction of waste materials: A facile review. *Heliyon, 9,* e12860.

Aigbe, U. O., Ukhurebor, K. E., & Onyancha, R. B. (Eds.). (2023). *Synthesis, characterization and applications of magnetic nanomaterials: Recent advances.* Springer Nature. https://doi.org/10.1007/978-3-031-36088-6

Aminot, Y., Sayfritz, S. J., Thomas, K. V, Godinho, L., Botteon, E., Ferrari, F., Boti, V., Albanis, T., Köck-Schulmeyer, M., & Diaz-Cruz, M. S. (2019). Environmental risks associated with contaminants of legacy and emerging concern at European aquaculture areas. *Environmental Pollution, 252,* 1301–1310.

Anani, A. A., Adama, K. K., Ukhurebor, K. E., Habib, A. H., Abanihi, V. K., & Pal, K. (2023). Application of nanofibrous protein for the purification of contaminated water as a next generational sorption technology: A review. *Nanotechnology, 34*(232004), 1–18.

Appleman, T. D., Higgins, C. P., Quinones, O., Vanderford, B. J., Kolstad, C., Zeigler-Holady, J. C., & Dickenson, E. R. (2014). Treatment of poly-and perfluoroalkyl substances in US full-scale water treatment systems. *Water Research, 51,* 246–255. https://doi.org/10.1016/j.watres.2013.10.067

Baabish, A., Sobhanei, S., & Fiedler, H. (2021). Priority perfluoroalkyl substances in surface waters-a snapshot survey from 22 developing countries. *Chemosphere, 273,* 129612.

Bai, X., & Son, Y. (2021). Perfluoroalkyl substances (PFAS) in surface water and sediments from two urban watersheds in Nevada, USA. *Science of the Total Environment, 751,* 141622.

Banzhaf, S., Filipovic, M., Lewis, J., Sparrenbom, C. J., & Barthel, R. (2017). A review of contamination of surface-, ground-, and drinking water in Sweden by perfluoroalkyl and polyfluoroalkyl substances (PFASs). *Ambio, 46*(3), 335–346.

Benbrahim-Tallaa, L., Lauby-Secretan, B. Loomis, D., Guyton, K. Z., Grosse, Y., Bouvard, F. El Ghissassi, V., Guha, N., Mattock, H., & Straif, K. (2014). Carcinogenicity of perfluorooctanoic acid, tetrafluoroethylene, dichloromethane, 1,2-dichloropropane, and 1,3-propane sultone. *The Lancet Oncology, 15*(9), 924-925. https://doi.org/10.1016/S1470-2045(14)70316-X

Birnbaum, L. S., & Grandjean, P. (2015). Alternatives to PFAS: Perspectives on the science. *Environmental Health Perspectives, 123*(5), A104–A105. https://doi.org/10.1289/ehp.1509944

Buck, R. C., Murphy, P. M., & Pabon, M. (2012). Chemistry, Properties, and Uses of Commercial Fluorinated Surfactants. In: Knepper, T., Lange, F. (eds) *Polyfluorinated Chemicals and Transformation Products*. The Handbook of Environmental Chemistry(), vol 17. Springer, Berlin, Heidelberg. https://doi.org/10.1007/978-3-642-21872-9_1

Buck, R. C., Franklin, J., Berger, U., Conder, J. M., Cousins, I. T., De Voogt, P., Jensen, A. A., Kannan, K., Mabury, S. A., & van Leeuwen, S. P. J. (2011). Perfluoroalkyl and polyfluoroalkyl substances in the environment: Terminology, classification, and origins. *Integrated Environmental Assessment and Management*, 7(4), 513–541.

Chen, C., Lu, Y., Zhang, X., Geng, J., Wang, T., Shi, Y., Hu, W., & Li, J. (2009). A review of spatial and temporal assessment of PFOS and PFOA contamination in China. *Chemistry and Ecology*, 25(3), 163–177.

Conder, J. M., Hoke, R. A., Wolf, W. D., Russell, M. H., & Buck, R. C. (2008). Are PFCAs bioaccumulative? A critical review and comparison with regulatory criteria and persistent lipophilic compounds. *Environmental Science & Technology*, 42(4), 995–1003. https://doi.org/10.1021/es070895g

Cookson, E. S., & Detwiler, R. L. (2022). Global patterns and temporal trends of perfluoroalkyl substances in municipal wastewater: A meta-analysis. *Water Research*, 221, 118784. https://doi.org/10.1016/j.watres.2022.118784

Corsini, E., Luebke, R. W., Germolec, D. R., & DeWitt, J. C. (2014). Perfluorinated compounds: Emerging POPs with potential immunotoxicity. *Toxicology Letters*, 230(2), 263–270.

Cousins, I. T., Johansson, J. H., Salter, M. E., Sha, B., & Scheringer, M. (2022). Outside the safe operating space of a new planetary boundary for per- and polyfluoroalkyl substances (PFAS). *Environmental Science & Technology*, 56, 11172–11179. https://doi.org/10.1021/acs.est.2c02765

Dasu, K., Xia, X., Siriwardena, D., Klupinski, T. P., & Seay, B (2022). Concentration profiles of per- and polyfluoroalkyl substances in major sources to the environment. *Journal of Environmental Management*, 301, 113879. https://doi.org/10.1016/j.jenvman.2021.113879

Dawson, D. E., Lau, C., Pradeep, P., Sayre, R. R., Judson, R. S., Tornero-Velez, R., & Wambaugh, J. F. (2023). A machine learning model to estimate toxicokinetic half-lives of per-and polyfluoro-alkyl substances (PFAS) in multiple species. *Toxics*, 11(2), 98. https://doi.org/10.3390/toxics11020098

Du, Z., Deng, S., Bei, Y., Huang, Q., Wang, B., Huang, J., & Yu, G. (2014). Adsorption behavior and mechanism of perfluorinated compounds on various adsorbents-A review. *Journal of Hazardous Materials*, 274, 443–454. https://doi.org/10.1016/j.jhazmat.2014.04.038

ECCC, HC. (2023a). *Draft state of per- and polyfluoroalkyl substances (PFAS) report*. Environment and Climate Change Canada and Health Canada. Available online: https://www.canada.ca/en/environment-climate-change/services/evaluating-existing-substances/draftstate-per-polyfluoroalkyl-substances-report.html (Accessed 20 May, 2025).

ECCC, HC. (2023b). *Risk management scope for per- and polyfluoroalkyl substances (PFAS)*. Environment and Climate Change Canada and Health Canada. Available online: https://www.canada.ca/en/environment-climate-change/services/evaluating-existing-substances/risk-management-scope-per-polyfluoroalkyl-substances.html (Accessed 20 May, 2025).

Ehsan, M. N., Riza, M., Pervez, N. Md., Omar Khyum, M. M., Liang, Y., & Naddeo, V. (2023). Environmental and health impacts of PFAS: Sources, distribution and sustainable management in North Carolina (USA). *Science of The Total Environment*, 878, 163123. https://doi.org/10.1016/j.scitotenv.2023.163123

Ehsan, M. N., Riza, M., Pervez, Md.P., Li, C-W., Zorpas, A.A., & Naddeo, V. (2024). PFAS contamination in soil and sediment: Contribution of sources and environmental impacts on soil biota. Case Studies in Chemical and Environmental Engineering, 9, 100643. https://doi.org/10.1016/j.cscee.2024.100643.

Ellis, D. A., Martin, J. W., De Silva, A. O., Marbury, S. A., Hurley, M. D., Sulbaek Andersen, M. P., & Wallington, T. J. (2004). Degradation of fluorotelomer alcohols: A likely atmospheric source of perfluoronated carboxylic acids. *Environmental Science & Technology*, *38*(12), 3316–3321. https://doi.org/10.1021/es049860w

Elmoznino, J., Vlahos, P., & Whitney, M. (2018). Occurrence and partitioning behavior of perfluoroalkyl acids in wastewater effluent discharging into the Long Island Sound. *Environmental Pollution*, *243*, 453–461. https://doi.org/10.1016/j.envpol.2018.07.076

EPA. (2014). U.S. Environmental Protection Agency. *Emerging contaminants fact sheet – Perfluorooctane Sulfonate (PFOS) and Perfluorooctanoic Acid (PFOA)*. EPA 505-F-14-001.

EPA. (2017). U.S. Environmental Protection Agency. *Technical fact sheet: Perfluorooctane Sulfonate (PFOS) and Perfluorooctanoic Acid (PFOA)*. EPA 505-F-17-001.

EPA. (2019a). U.S. Environmental Protection Agency. Method 537.1 Determination of selected per- and polyfluorinated alkyl substances in drinking water by solid phase extraction and liquid chromatography/tandem mass spectrometry (LC/MS/MS). Available online: https://cfpub.epa.gov/si/si_public_record_Report.cfm?dirEntryId =343042&Lab=NERL (Accessed 20 May, 2025).

EPA. (2019b). U.S. Environmental Protection Agency. SW-486 update VII announcements, phase II – PFAS 8372 and 3512. Available online: https://www.epa.gov/hw-sw846/sw -846-update-vii-announcements (Accessed 20 May, 2025).

Espana, V. A. A., Mallavarapu, M., & Naidu, R. (2015). Treatment technologies for aqueous perfluorooctanesulfonate (PFOS) and perfluorooctanoate (PFOA): A critical review with an emphasis on field testing. *Environmental Technology & Innovation*, *4*, 168–181.

Fenton, S. E., Ducatman, A., Boobis, A., DeWitt, J. C., Lau, C., Ng, C., Smith, J. S., & Roberts, S. M. (2021). Per- and polyfluoroalkyl substance toxicity and human health review: Current state of knowledge and strategies for informing future research. Environmental Toxicology and Chemistry, *40*, 606–630. https://doi.org/10.1002/etc.4890

Fernandez, N. A., Rodriguez-Freire, L., Keswani, M., & Sierra-Alvarez, R. (2016). Effect of chemical structure on the sonochemical degradation of perfluoroalkyl and polyfluo- roalkyl substances (PFASs). *Environmental Science: Water Research & Technology*, *2*(6), 975–983.

Ferrey, M. L., Wilson, J. T., Adair, C., Su, C., Fine, D. D., Liu, X., & Washington, J. W. (2012). Behavior and fate of PFOA and PFOS in sandy aquifer sediment. *Groundwater Monitoring & Remediation*, *32*(4), 63–71. https://doi.org/10.1111/j.1745-6592.2012 .01395.x

Gallen, C., Bignert, A., Taucare, G., O'Brien, J., Braeunig, J., Reeks, T., Thompson, J., & Mueller, J. F. (2022). Temporal trends of perfluoroalkyl substances in an Australian wastewater treatment plant: A ten-year retrospective investigation. *Science of the Total Environment*, 150211. https://doi.org/10.1016/j.scitotenv.2021.150211

Gebbink, W. A., & van Leeuwen, S. P. J. (2020). Environmental contamination and human exposure to PFASs near a fluorochemical production plant: Review of historic and cur- rent PFOA and GenX contamination in the Netherlands. *Environment International*, *137*, 105583.

Gewurtz, S. B., Auyeung, A. S., De Silva, A. O., Teslic, S., & Smyth, S. A. (2024). Per- and polyfluoroalkyl substances (PFAS) in Canadian municipal wastewater and biosolids: Recent patterns and time trends 2009 to 2021. *Science of the Total Environment*, *912*, 168638. https://doi.org/10.1016/j.scitotenv.2023.168638

Ghisi, R., Vamerali, T., & Manzetti, S. (2019). Accumulation of perfluorinated alkyl substances (PFAS) in agricultural plants: A review. *Environmental Research*, *169*, 326–341.

Glüge, J., Scheringer, M., Cousins, I. T., DeWitt, J. C., Goldenman, G., Herzke, D., Lohmann,
R., Ng, C. A., Trier, X., & Wang, Z. (2020). An overview of the uses of per-and polyfluoroalkyl substances (PFAS). *Environmental Science: Processes & Impacts*, *22*(12),
2345–2373.

Goodrow, S. M., Ruppel, B., Lippincott, R. L., Post, G. B., & Procopio, N. A. (2020).
Investigation of levels of perfluoroalkyl substances in surface water, sediment and fish
tissue in New Jersey, USA. *Science of the Total Environment*, *729*, 138839.

Gottschall, N., Topp, E., Edwards, M., Payne, M., Kleywegt, S., & Lapen, D. R. (2017).
Brominated flame retardants and perfluoroalkyl acids in groundwater, tile drainage,
soil, and crop grain following a high application of municipal biosolids to a field.
Science of the Total Environment, *574*, 1345–1359. https://doi.org/10.1016/j.scitotenv
.2016.08.044

Grandjean, P., Andersen, E. W., Budtz-Jørgensen, E., Nielsen, F., Mølbak, K., Weihe, P., &
Heilmann, C. (2012). Serum vaccine antibody concentrations in children exposed to
perfluorinated compounds. *Jama*, *307*(4), 391–397.

Harding-Marjanovic, K. C., Houtz, E. F., Yi, S., Field, J. A., Sedlak, D. L., & Alvarez-
Cohen, L. (2015). Aerobic biotransformation of fluorotelomer thioether amido sulfonate (Lodyne) in AFFF-amended microcosms. *Environmental Science & Technology*,
49(13), 7666–7674. https://doi.org/10.1021/acs.est.5b01219

Higgins, C. P., & Luthy, R. G. (2006). Sorption of perfluorinated surfactants on sediments.
Environmental Science & Technology, *40*(23), 7251–7256. https://doi.org/10.1021/
es061000

Houde, M., De Silva, A. O., Muir, D. C. G., & Letcher, R. J. (2011). Monitoring of perfluorinated compounds in aquatic biota: An updated review: PFCs in aquatic biota.
Environmental Science & Technology, *45*(19), 7962–7973.

Houtz, E. F., Higgins, C. P., Field, J. A., & Sedlak, D. L. (2013). Persistence of perfluoroalkyl
acid precursors in AFFF-impacted groundwater and soil. *Environmental Science &
Technology*, *47*(15), 8187–8195. https://doi.org/10.1021/es4018877

Huang, S., & Jaffe, P. R. (2019). Defluorination of perfluorooctanoic acid (PFOA) and perfluorooctane sulfonate (PFOS) by Acidimicrobium sp. strain A6. *Environmental Science
and Technology*, *53*, 11410–11419. https://doi.org/10.1021/acs.est.9b04047

IARC. (2016). *International Agency for Research on Cancer. Monographs on the evaluation
of carcinogenic risks to humans. Lists of classifications, volumes 1 to 116.* IARC.

ITRC. (2018). Interstate Technology and Regulatory Council. Naming conventions and physical and chemical properties of per- and polyfluoroalkyl substances (PFAS). Available
online: https://pfas-1.itrcweb.org/wp-content/uploads/2018/03/pfas_fact_sheet_nam-
ing_conventions__3_16_18.pdf (Accessed 20 May, 2025).

ITRC. (2020). *Naming conventions for per- and polyfluoroalkyl substances (PFAS).* Available
online: https://pfas-dev.itrcweb.org/wp-content/uploads/2020/10/naming_conventions
_508_2020Aug_Final.pdf (19 May, 2025).

ITRC. (2022). *PFAS technical and regulatory guidance document and fact sheets.* Interstate
Technology & Regulatory Council. Available online: https://pfas-1.itrcweb.org (19
May, 2025).

Jian, J.-M., Guo, Y., Zeng, L., Liang-Ying, L., Lu, X., Wang, F., & Zeng, E. Y. (2017). Global
distribution of perfluorochemicals (PFCs) in potential human exposure source–a
review. *Environment International*, *108*, 51–62.

Johnson, G. R. (2022). PFAS in soil and groundwater following historical land application
of biosolids. *Water Research*, *211*, 118035. https://doi.org/10.1016/j.watres.2021.118035

Johnson, R. L., Anschutz, A. J., Smolen, J. M., Simcik, M. F., & Penn, R. L. (2007). The adsorption of perfluorooctane sulfonate onto sand, clay, and iron oxide surfaces. *Journal of Chemical & Engineering Data, 52*(4), 1165–1170. https://doi.org/10.1021/je060285g

Kerry, R. G., Montalbo, F. J. P., Das, R., Patra, S., Mahapatra, G. P., Maurya, G. K., Nayak, V., Jena, A. B., Ukhurebor, K. E., Jena, R. C., Gouda, S., Majhi, S., & Rou, J. R. (2022). An overview of remote monitoring methods in biodiversity conservation. *Environmental Science and Pollution Research, 2022*, 1–43.

Kucharzyk, K. H., Darlington, R., Benotti, M., Deeb, R., & Hawley, E. (2017). Novel treatment technologies for PFAS compounds: A critical review. *Journal of Environmental Management, 204*, 757–764.

Li, F., Zhang, C., Qu, Y., Chen, J., Chen, L., Liu, Y., & Zhou, Q. (2010). Quantitative characterization of short-and long-chain perfluorinated acids in solid matrices in Shanghai, China. *Science of the Total Environment, 408*(3), 617–623.

Mahinroosta, R., & Senevirathna, L. (2020). A review of the emerging treatment technologies for PFAS contaminated soils. *Journal of Environmental Management, 255*, 109896.

MDH. (2018). Minnesota Department of Health, 2018. Media FAQ: Fish Consumption Advisory, PFOS and Lake Elmo Fish 2018. Available online: https://www.co.washington.mn.us/DocumentCenter/View/20895/FAQ-2018-Fish-Consumption-Advisory-NR (Accessed 20 May, 2025).

Nwankwo, W., Adetunji, C. O., Ukhurebor, K. E., Acheme, D. I., Makinde, S. A., Nwankwo, C. P., & Umezuruike, C. (2023). Sector-independent integrated system architecture for profiling hazardous industrial wastes. In Z. Hu, I. Dychka, & M. He (Eds.), *Advances in computer science for engineering and education VI. International Conference on Computer Science, Engineering and Education Applications (ICCSEEA) 2023.* Lecture Notes on Data Engineering and Communications Technologies, Springer, Cham, 181, 721–747.

Obodo, K. (2024). The application of magnetic sorbents for the sequestration of pesticides, pharmaceuticals, and perfluoroalkyl and polyfluoroalkyl substances from aqueous solutions. In K. E. Ukhurebor & U. O. Aigbe (Eds.), *Environmental applications of magnetic sorbents* (pp. 6–1 to 1–18). Institute of Physics Publishing. https://doi.org/10.1088/978-0-7503-5909-2ch6.

OECD. (2015). Organisation for Economic Cooperation and Development. Working Towards a Global Emission Inventory of PFASs: Focus on PFCAs – Status Quo and the Way Forward. Paris: Environment, Health and Safety, Environmental Directorate, OECD/UNEP Global PFC Group. Available online: https://www.oecd.org/en/publications/working-towards-a-global-emission-inventory-of-pfass-focus-on-pfcas-status-quo-and-the-way-forward_f97f34b1-en.html (19 May, 2025).

Oliaei, F., Kriens, D., Weber, R., & Watson, A. (2013). PFOS and PFC releases and associated pollution from a PFC production plant in Minnesota (USA). *Environmental Science and Pollution Research, 20*, 1977–1992.

Parsons, J. R., Sáez, M., Dolfing, J., & De Voogt, P. (2008). Biodegradation of perfluorinated compounds. *Reviews of Environmental Contamination and Toxicology, 196*, 53–71.

Quinete, N., Wu, Q., Zhang, T., Yun, S. H., Moreira, I., & Kannan, K. (2009). Specific profiles of perfluorinated compounds in surface and drinking waters and accumulation in mussels, fish, and dolphins from southeastern Brazil. *Chemosphere, 77*(6), 863–869.

Quinete, N. S., & Ogunbiyi, O. D. (2024). PFAS 'forever chemicals' are getting into ocean ecosystems, where dolphins, fish and manatees dine – we traced their origins. The Conversation, Published: November 14, 2023 3.25pm SAST Updated: April 10, 2024,

7.51 pm SAST. Available online: https://theconversation.com/pfas-forever-chemi-cals-are-getting-into-ocean-ecosystems-where-dolphins-fish-and-manatees-dine-we-traced-their-origins-216254.

Rahman, M. F., Peldszus, S., & Anderson, W. B. (2014). Behaviour and fate of perfluoroalkyl and polyfluoroalkyl substances (PFASs) in drinking water treatment: A review. *Water Research, 50*, 318–340.

Rayne, S., & Forest, K. (2009). Perfluoroalkyl sulfonic and carboxylic acids: A critical review of physicochemical properties, levels and patterns in waters and wastewaters, and treatment methods. *Journal of Environmental Science and Health Part A, 44*(12), 1145–1199.

Redmon, J. H., DeLuca, N. M., Thorp, E., Liyanapatirana, C., Allen, L., Andrew J., & Kondash, A. J. (2025). Hold my beer: The linkage between municipal water and brewing location on PFAS in popular beverages. *Environmental Science & Technology, 59*(17), 8368–8379. https://doi.org/10.1021/acs.est.4c11265

Schulz, K., Silva, M. R., & Klaper, R. (2020). Distribution and effects of branched versus linear isomers of PFOA, PFOS, and PFHxS: A review of recent literature. *Science of the Total Environment, 733*, 139186.

Sharifan, H., Bagheri, M., Wang, D., Burken, J. G., Higgins, C. P., Liang, Y., Liu, J., Schaefer, C. E., & Blotevogel, J. (2021). Fate and transport of per-and polyfluoroalkyl substances (PFASs) in the vadose zone. *Science of the Total Environment, 771*, 145427.

Sinclair, E., Mayack, D. T., Roblee, K., Yamashita, N., & Kannan, K. (2006). Occurrence of perfluoroalkyl surfactants in water, fish, and birds from New York State. *Archives of Environmental Contamination and Toxicology, 50*(3), 398–410. https://doi.org/10.1007/s00244-005-1188-z

Singh, R. P., Ukhurebor, K. E., Singh, J., Adetunji, C. O., & Singh, K. R. B. (Eds.). (2022). *Nanobiosensors for environmental monitoring - Fundamentals and application.* Springer Nature. https://doi.org/10.1007/978-3-031-16106-3

Steinle-Darling, E., & Reinhard, M. (2008). Nanofiltration for trace organic contaminant removal: Structure, solution, and membrane fouling effects on the rejection of perfluorochemicals. *Environmental Science & Technology, 42*(14), 5292–5297. https://doi.org/10.1021/es703207s

Sunderland, E. M., Hu, X. C., Dassuncao, C., Tokranov, A. K., Wagner, C. C., & Allen, J. G. (2019). A review of the pathways of human exposure to poly-and perfluoroalkyl substances (PFASs) and present understanding of health effects. *Journal of Exposure Science & Environmental Epidemiology, 29*(2), 131–147.

Tseng, N., Wang, N., Szostek, B., & Mahendra, S. (2014). Biotransformation of 6: 2 fluorotelomer alcohol (6: 2 FTOH) by a wood-rotting fungus. *Environmental Science & Technology, 48*(7), 4012–4020. https://doi.org/10.1021/es4057483

Ukhurebor, K. E., & Aigbe, U. O. (Eds.). (2024). *Environmental applications of magnetic sorbents.* Institute of Physics (IOP) Publishing. https://doi.org/10.1088/978-0-7503-5909-2

Ukhurebor, K. E., Aigbe, U. O., & Onyancha, R. B. (Eds.). (2023a). *Adsorption applications for environmental sustainability.* Institute of Physics (IOP) Publishing. https://doi.org/10.1088/978-0-7503-5598-8

Ukhurebor, K. E., Aigbe, U. O., Onyancha, R. B., Ndunagu, J. N., Osibote, O. A., Emegha, J. O., Balogun, V. A., Kusuma, H. S., & Darmokoesoemo, H. (2022). An overview on the emergence and challenges of land reclamation: Issues and prospect. *Applied and Environmental Soil Science, 5889823*, 1–14.

Ukhurebor, K. E., Aigbe, U. O., Onyancha, R. B., Hussain, A., Okundaye, B., Aidonojie, P. A., Siloko, B. E., Hossain, I., Kusuma, H. S., & Darmokoesoemo, H. (2024). Environmental influence of gas flaring: Perspective from the Niger Delta region of Nigeria. *Geofluids, 1321022*, 1–17.

Ukhurebor, K. E., Hossain, I., Pal, K., Jokthan, G., Osang, F., Ebrima, F., & Katal, D. (2023b). Applications and contemporary issues with adsorption for water monitoring and remediation: A facile review. *Topics in Catalysis*, , 66(9–10), 682–704.

Vo, H. N. P., Ngo, H. H., Guo, W., Nguyen, T. M. H., Li, J., Liang, H., Deng, L., Chen, Z., & Nguyen, T. A. H. (2020). Poly-and perfluoroalkyl substances in water and wastewater: A comprehensive review from sources to remediation. *Journal of Water Process Engineering, 36*, 101393.

Wang, S., Ma, L., Chen, C., Li, Y., Wu, Y., Liu, Y., Dou, Z., Yamazaki, E., Yamashita, N., & Lin, B.-L. (2020a). Occurrence and partitioning behavior of per-and polyfluoroalkyl substances (PFASs) in water and sediment from the Jiulong Estuary-Xiamen Bay, China. *Chemosphere, 238*, 124578.

Wang, T., Wang, P., Meng, J., Liu, S., Lu, Y., Khim, J. S., & Giesy, J. P. (2015). A review of sources, multimedia distribution and health risks of perfluoroalkyl acids (PFAAs) in China. *Chemosphere, 129*, 87–99.

Wang, W., Rhodes, G., Ge, J., Yu, X., & Li, H. (2020b). Uptake and accumulation of per-and polyfluoroalkyl substances in plants. *Chemosphere, 261*, 127584.

Wanninayake, D. M. (2021). Comparison of currently available PFAS remediation technologies in water: A review. *Journal of Environmental Management, 283*, 111977.

Winchell, L. J., Wells, M. J. M., Ross, J. J., Fonoll, X., Norton Jr, J. W., Kuplicki, S., Khan, M., & Bell, K. Y. (2021). Analyses of per-and polyfluoroalkyl substances (PFAS) through the urban water cycle: Toward achieving an integrated analytical workflow across aqueous, solid, and gaseous matrices in water and wastewater treatment. *Science of The Total Environment, 774*, 145257.

Xiao, F. (2017). Emerging poly-and perfluoroalkyl substances in the aquatic environment: A review of current literature. *Water Research, 124*, 482–495.

Yao, J., Pan, Y., Sheng, N., Su, Z., Guo, Y., Wang, J., & Dai, J. (2020). Novel perfluoroalkyl ether carboxylic acids (PFECAs) and sulfonic acids (PFESAs): Occurrence and association with serum biochemical parameters in residents living near a fluorochemical plant in China. *Environmental Science & Technology, 54*(21), 13389–13398.

Young, C. J., Furdui, V. I., Franklin, J., Koerner, R. M., Muir, D. C., & Mabury, S. A. (2007). Perfluorinated acids in arctic snow: New evidence for atmospheric formation. *Environmental Science & Technology, 41*(10), 3455–3461. https://doi.org/10.1021/es0626234

Zhang, D. Q., Zhang, W. L., & Liang, Y. N. (2019a). Adsorption of perfluoroalkyl and polyfluoroalkyl substances (PFASs) from aqueous solution-A review. *Science of the Total Environment, 694*, 133606.

Zhang, W., Cao, H., & Liang, Y. (2021a). Plant uptake and soil fractionation of five ether-PFAS in plant-soil systems. *Science of The Total Environment, 771*, 144805.

Zhang, W., Zhang, D., & Liang, Y. (2019b). Nanotechnology in remediation of water contaminated by poly-and perfluoroalkyl substances: A review. *Environmental Pollution, 247*, 266–276.

Zhang, Z., Sarkar, D., Datta, R., & Deng, Y. (2021b). Adsorption of perfluorooctanoic acid (PFOA) and perfluorooctanesulfonic acid (PFOS) by aluminum-based drinking water treatment residuals. *Journal of Hazardous Materials Letters, 2*, 100034.

2 Sources and Production of Perfluoroalkyl and Polyfluoroalkyl Substances

2.1 INTRODUCTION

Per/polyfluoroalkyl substances (PPFAS) consist of a various class of fluorinated artificial compounds considered by the existence of at least "perfluorinated methyl-groups $(-CF_3)$" or a "perfluorinated methylene-group $(-CF_2-)$", an adjustable amount of carbon (C) atoms, fluorination-degree and the existence of other chemical groups (Panieri, Baralic, Djukic-Cosic, Buha Djordjevic, & Saso, 2022; Obodo, 2024; Wang, DeWitt, Higgins, & Cousins, 2017). PPFAS are a huge collection of man-made/artificial chemicals that have been manufactured and utilized in industry and end-user products since as far back as the 1940s. Up to the present time, there are more than 1.4×10^3 categories of PPFAS. The most studied compounds amongst PPFAS are perfluorooctanoic acid (PFOAA) and perfluorooctane sulfonic acid (PFOSA), which are perfluoroalkyl acids (PFAAs) owing to their noxiousness and persistence in the ecosystem and the human body. Expressively highlighting their hazard to the environment, PFOAA and PFOSA were listed in the 2009 and 2019 "Stockholm Convention" as Persistent Organic Contaminants. They have been considered expansively global by academics since the 1980s. (Tang, Hamid, Yusoff, & Chan, 2023; Bai & Son, 2021; Davidson & Boland, 2023; Ehsan et al., 2024).

Many PPFAS have been found far and wide in different environmental matrices such as surface water (SW), soils, groundwater (GW), landfill effluents, wastewater (WW), and sediments. There is a growing fear by scientists and regulators about PPFAS' existence, persistence, and fate, owing to most approaches not being efficient in sequestrating short-chain (SC) PPFAS and being resistant to chemical and biological oxidation. Thus, an improved approach for recognizing and measuring PPFAS in different environments is urgently required (Bai & Son, 2021).

Though there are thousands of diverse PPFAS, only a few are being observed and have supervisory concerns like PFOAA, PFOSA, perfluorohexane sulfonate acid (PFHxSA), perfluorononanoic acid (PFNAA), and perfluorodecanoic acid (PFDA). PFOSA and PFOAA are the most distinguished and discovered PPFAS in the universal bionetwork, and while they have been circumscribed in current years, numerous alternatives have improved considerably as displayed in Figure 2.1. Current study

DOI: 10.1201/9781003625537-2

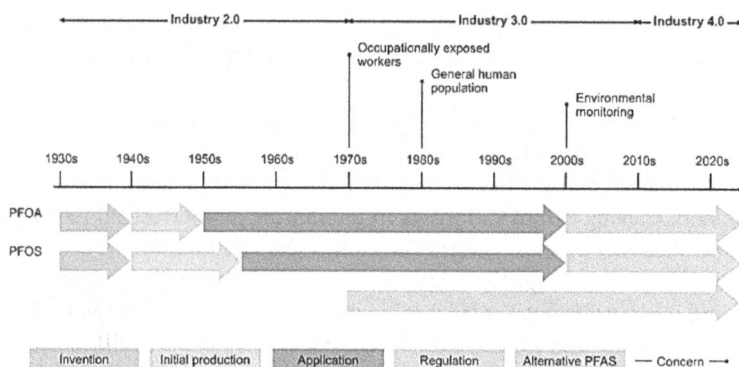

FIGURE 2.1 A sequence of events of PPFAS in relationship to their origination, manufacturing, uses, health and monitoring concerns, and guidelines. Adapted and reproduced from Perez et al. (2023) with permission from Elsevier B.V.

and monitoring, health impacts, technology of treatment, and guidelines remain inadequate for PFOSA and PFOAA, with some physiognomies, modes, factors, and mechanisms hitherto to be studied, as well as persistence, build up in humans, animals, and the ecosystem, exposure in a diversity of forms and possibly of environmental and human-health (HH) impacts remains a huge concern.

In human tissue, PPFAS has been found to bioaccumulate, and this bioaccumulation can lead to long-lasting diseases of choice organs like liver damage, reduced fertility, obesity, thyroid disease, cancer, high cholesterol, and hormone suppression. These effects can be predominantly severe in susceptible populations like pregnant women, newborns, and persons with compromised immune systems (Sabba et al., 2025).

The unreceptive impacts of PPFAS on people are becoming a serious issue internationally, owing to their extended period of persistence, curtailed biodegradability, and elevated bioaccumulation in humans, as well as flora and fauna. They are ideal for use in paper products, packaging of food, non-stick cookware, water-soluble film-forming foams employed in firefighting and textile coatings owing to their exclusive physical and chemical properties like extreme-thermal- and chemical-stability and water- and oil-repellency. The conditional health recommended value for PFOAA and PFOSA of 0.4-μg.L^{-1} and 0.2-μg.L^{-1} was established by the EPA in 2009. A lifetime health recommendation for long-term exposure of 0.07 μg.L^{-1} to PFOSA and PFOAA through drinking water (DW) was issued in 2016. Among the several PPFAS, perfluoroalkyl carboxylic-acids (PFCAs) and perfluoroalkyl sulfonic acids (PFSAs) are sturdy acids with pKa projected to be near zero for PFCAs and -1 for PFSAs, which shows that they exist in ionic forms in most ecological settings (Bai & Son, 2021).

PPFAS' common physiognomies are epitomized by their chemical stability, which causes ecological persistence and their preeminent-mobility, and this confers in them an extended array of transport-potential, triggering their prevalent distribution even into isolated areas such as the Arctic or Antarctic, and their tendency to bioaccumulate and biomagnify in biota via food chain contamination (Panieri, Baralic, Djukic-Cosic, Buha Djordjevic, & Saso, 2022; Wang et al., 2021; Kurwadkar et al., 2022). The existence of some PPFAS has been reported in milk, tissues, organs, blood, and urine of various human residents living in industrialized nations and has been linked to several adverse health impacts. Also, pertinent PPFAS concentrations have been sensed in GW, freshwater, seawater, air, soil, and DW, possibly causing ecotoxic impacts in the marine and terrestrial bionetworks at the trophic phases of main producers and consumers and subordinate consumers. An extra film of intricacy is specified by the co-occurrence of various combinations of PPFAS and other pollutants in the ecological medium, for which quantifiable hazard evaluation study and toxicologic or ecotoxicologic data are still unusual, if not absent. Thus, it is of extreme standing to swiftly fill the inconsistency of systematic information and to create acceptable systematic approaches to detect PPFAS pollution and the high throughput methods to forecast PPFAS noxiousness and efficient strategies of statutory guidelines, remediation, and therapeutic interventions to moderate possible impacts on biota and humans. It is also expected that the government and industry would collaborate to increasingly encourage a more justifiable innovation and a change in the direction of less unsafe substances by delivering proper strategies or supervisory actions to control and monitor PPFAS ecological release (Panieri, Baralic, Djukic-Cosic, Buha Djordjevic, & Saso, 2022; Kurwadkar et al., 2022; Gaines, 2023).

2.2 SOURCES, PROPERTIES, AND USES OF PPFAS AND THEIR ENVIRONMENTAL SEGMENTS DISTRIBUTION

2.2.1 CLASSIFICATION OF PPFAS

As the forever compounds family swiftly grows, it is almost impossible to establish threat data linked with individual novel PPFAS chemistry. Hence, having an expressive grouping of PPFAS compounds is tremendously significant (Su & Rajan, 2021). From a historic viewpoint, the foremost categorization of PPFAS was projected around 2011 by Buck, et al. in a seminal article, where PPFAS were well-defined as "the extremely fluorinated-aliphatic substances that comprise of one or more C atoms on which all the hydrogen (H) proxies have been substituted by fluorine (F) atoms, in such a way that they consist of the perfluoroalkyl-moiety C_nF2n+1-". In 2018, the existence of numerous PPFAS molecules was reported by the Organization for Economic Co-operation and Development (OECD), regardless of them having completely fluorinated C atoms, they were devoid of the $-CF_3$ group and, hence, did not meet the previous definition by Buck, et al. A comprehensive classification of PPFAS was proposed in a current report by OECD as "fluorinated substances that comprises of at least a fully fluorinated methyl and methylene C atom (deprived of the attachment of any H/chlorine (Cl)/bromine (Br)/iodine (I) atoms) such that, with a few noted exemptions (characterized by the attachment of a C atom as a substitute than having H/Cl/Br/I atoms), any chemical with at least a $-CF_3$ or $-CF_2$." (Panieri, Baralic, Djukic-Cosic, Buha Djordjevic, & Saso, 2022; Buck et al., 2011; Buck,

Korzeniowski, Laganis, & Adamsky, 2021; Wang et al., 2021; Schymanski et al., 2023). Globally, approximately 4.3×10^3 PPFAS are listed in the Chemical Abstracts Service. Various categorizations of PPFAS exist, but a modest draft outline is mentioned in Figure 2.2 (Ambaye, Vaccari, Prasad, & Rtimi, 2022).

Regularly utilized as a key discriminant and an excellent forecaster of physico-chemical properties, bioaccumulation, protein binding, and ecological fate distribution is the length of the fluorinated C-chain, which varies between C4 and C17, and is applied for PPFAS classification. They can be grouped into non-polymeric (low-molecular weight) and polymeric (high-molecular weight) molecules. In the non-polymeric PPFAS, as displayed in Figure 2.3, it comprises of molecules in which the hydrophobic (HYPIC) C-chain is completely fluorinated with the exclusion of the terminal-end, which hosts polar-functional-groups (PFGs) like carboxylate (COO-), phosphate (OPO_3^-) or sulfonate (SO_3^-), which bestows hydrophilicity. The perfluoro-alkyl PPFAS can be additionally segmented into diverse subcategories as displayed in Table 2.1, among which are PFAAs, which consist of the most well-recognized and broadly researched molecules like PFOSA and PFOAA, as shown in Figure 2.4 (a and d). By distinction, the polyfluoroalkyl PPFAS group incorporates molecules in which at least one, but not all, of the C atoms are in part fluorinated and bound to oxygen (O) to H atoms (Panieri, Baralic, Djukic-Cosic, Buha Djordjevic, & Saso, 2022; Brunn et al., 2023). These substances/molecules are more mobile and reactive,

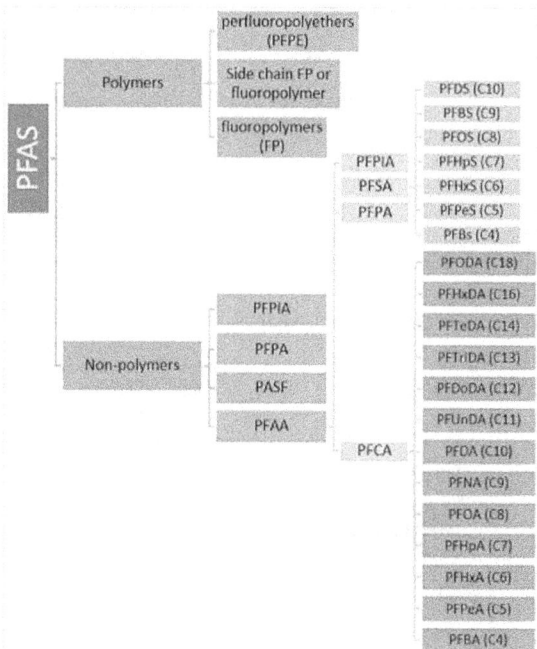

FIGURE 2.2 Cataloguing of PPFAS. Adapted and reproduced from Ambaye et al. (2022) with permission from Springer Nature.

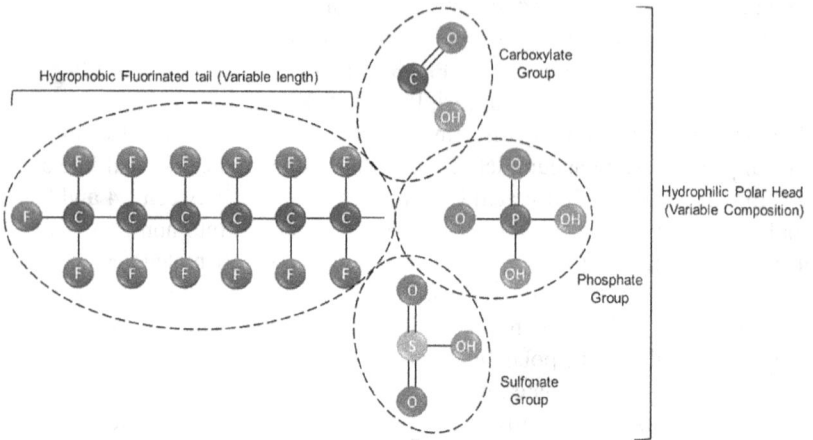

FIGURE 2.3 Outline of the wide-ranging structure of non-polymeric, perfluorinated PPFAS substances. Adapted and reproduced from Panieri et al. (2022) with permission from MDPI, Basel, Switzerland open access article distributed under the terms and conditions of the Creative Commons Attribution (CC BY) license.

TABLE 2.1
Classification of PPFAS based on their categories and subcategories

Non-polymeric PPFAS		Polymeric PPFAS
Perfluorinated PPFAS	**Polyfluorinated PPFAS**	**Perfluorinated PPFAS**
Subcategory		
PFAAs, PFSAs, perfluoroalkane sulfonic (PFSIAs), perfluorocarboxylic acids, perfluoroalkyl phosphonic acids (PFPAs), and perfluoroalkyl phosphinic acids (PFPIAs)	Fluorotelomer compounds (FT)	FPrs
Perfluoroalkyl ether acids (PFEAs)	Perfluoroalkane sulfonamido compounds	Side-chain FPs
Perfluoroalkane sulphonamides (FASA)		Perfluoropolyethers (PFPEs)
Perfluoroalkane sulfonyl Fs (PASFs)		
Perfluoroalkyl iodides (PFAIs)		
Perfluoroalkyl Fs (PAFs)		
Perfluoroalkyl aldehydes (PFALs)		

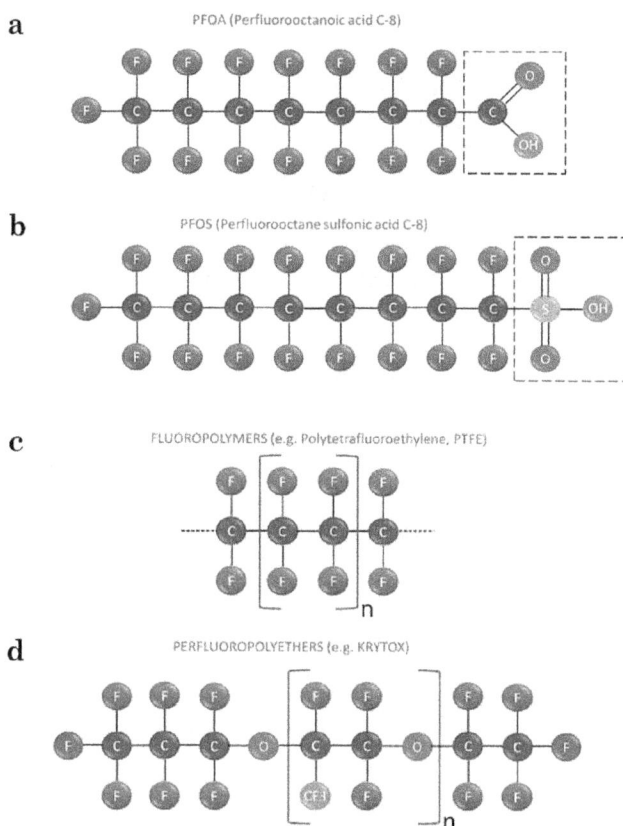

FIGURE 2.4 Description of Non-polymeric and polymeric instances (a and b) PFOAA and PFOSA, (c) PTFE structure (Teflon), a polymeric PPFAS belonging to the subcategory of FPrs are established by a moiety of $(CF2-CF2)_n$ atoms repetitive up thousands of times and structure of a lubricant (Krytox), typically a polymeric PPFAS belonging to the subcategory of the perfluoropolyethers and "$(CF[CF3]-CF2-O)_n$ moiety" is recurring among 10 to 60 times. Adapted and reproduced from Panieri et al. (2022) with permission from MDPI, Basel, Switzerland open access article distributed under the terms and conditions of the Creative Commons Attribution (CC BY) license.

hence, voluntarily transported after exposure in wildlife and human body tissue. They are synthesized by blending various undistinguishable monomers or trivial molecules (Ambaye, Vaccari, Prasad, & Rtimi, 2022; Xu et al., 2021).

Polymeric fluoropolymers (FPrs), fluorinated polymers (FPs), and perfluoropoly-ether are key subcategories of PPFAS and can be degraded into terminal PFAA products (Ambaye, Vaccari, Prasad, & Rtimi, 2022). These polymeric PPFAS, such as FPrs substances, where most or else all H atoms of the C-chains are substituted by fluoride (F⁻) atoms, such as polytetrafluoroethylene (PTFE) and Polyvinylidene Fluoride (PVDF); side-chain FPs, these substances are established by non-fluori-nated C-chains of adaptable structure and poly or perfluoroalkylic side chains such as fluorinated acrylate polymers and perfluoropolyethers (PFPEs) are substances where the key backbone comprise of O atoms and F- atoms are precisely bounded to the C-chain as shown in Figure 2.4c and d and Table 2.1 (Panieri, Baralic, Djukic-Cosic, Buha Djordjevic, & Saso, 2022).

2.3   PFAS PRODUCTION APPROACHES

PPFAS industrial production originated with the 3M Company in 1930, and it was further advanced by the DuPont Company to create the PTFE, which was viably traded as Teflon with extensive uses as an unchanging, non-reactive polymer active in various applications as defined in Figure 2.5. Owing to the existence of very stable C-F bonds, PPFAS have raised chemical and thermal stability, water, heat, grease, and stain resistance, and a tremendously diminutive friction-coefficient. Specifically, PPFAS are lipophobic and HYPIC, thus their status for application as non-stick glazes. They are also applied in the military, aerospace, paints, pesticides, leather industries, insulation materials in electronics, glazes for truncated friction bearings, and several customer products such as non-stick cookware, food packaging material, drinks, pesticides, and paint can line materials. For an improved consideration of the ecological existence and behaviour of PPFAS and the associations between the PPFAS families, it is advantageous to define the two principal production processes employed to create compounds comprising of PPFAS chains. Regularly, the two key methods that have been utilized to produce PPFAS and small-chain FPs are telomerization and electrochemical fluorination (ECF) (Dhore & Murthy, 2021; Buck et al., 2011).

2.3.1   ELECTROCHEMICAL FLUORINATION (ECF)

ECF was established by Joseph Simons, an American chemist, with the 3M Company's backing. This approach was applied for creating the perfluorinated carboxylic acids, amines, sulfonic acids, and others. This involves the organic compound dispersion in anhydrous HF- and the passing of direct current (DC) into the

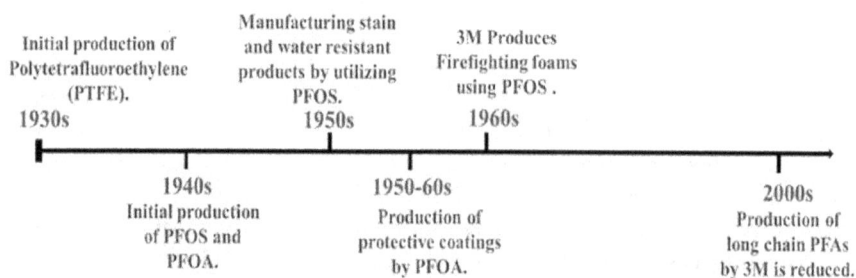

FIGURE 2.5   Manufacture and usage of PPFAS. Adapted and reproduced from Dhore & Murthy (2021) with permission from Elsevier B.V.

obtained solution, which results in the H atom replacement with F in an organic compound. The ECF creates a blend of C-chain lengths with odd and even numbers, with an uneven 20–30% and 70–80% branched and linear substances in the instance of PFOSA and PFOAA synthesis (Dhore & Murthy, 2021; Buck et al., 2011).

2.3.2 TELOMERIZATION (TLN)

The TLN approach is an essential manufacturing process for creating PPFAAs, which is more multifaceted and problematic to operate than the electrochemical fluoridation but creates an ultimate product with elevated purity and unconventional yields. It is generally applied in the creation of fluorotelomer substances (acrylates [FTAs], fluoro-telomer olefins [FTOs], and fluoro-telomer alcohols [FTOHs]). Using this process, PPFAS with a direct chain with minute or no branching is created, and this process mostly yields even-numbered, straight C-chain isomers. At a worldwide scale, an alternative approach for the manufacturing production of PPFAS is polymerization, which is chiefly known for producing fluoro-polymers, particularly PTFE. TLN is categorized as the polymerization reaction between a polymerizable species M (taxogen) and a compound XY (telogen) as illustrated by **Equation 2.1** (Dhore & Murthy, 2021; Buck et al., 2011).

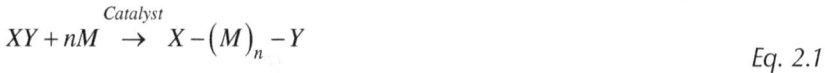

$$XY + nM \xrightarrow{Catalyst} X-\left(M\right)_n-Y$$

Eq. 2.1

2.4 SOURCES AND TRANSPORT PATHWAYS

The worldwide pollution by PPFAS is an outcome of an extensive array of industrial and consumer-production, manufacturing, and applications. This pollution arises from both principal industrial-facilities, which create and use PPFAS to create goods and produce PPFAS unrestricted to the environment via emission, dispersion, spill and waste disposal, and effluents (Wee & Aris, 2023; Sunderland et al., 2019). The important PPFAS pollution sources and their probable ecological sections are point- and non-point sources, which are shown in Figure 2.6. The point-based sources are fixed and distinct, and these comprise of firefighting training places, effluent treatment components, landfills, manufacturing and industrial facilities, etc. The non-point-based sources consist of diffuse sources of unknown sites or sources, and these include surface overflow, atmospheric imposition of volatile PPFAS, precipitation, degradation precursor compounds, etc. PPFAS include, in cooperation, HYPIC (fluorinated C-chain) and hydrophilic (FGs) sections. But PPFAS solubility is reliant on the functional moieties and the length of the chains, which also controls their existence in the bionetwork. Longer chains of PPFAS are slightly water-soluble and vice-versa. While SCs of PPFAS (C ≤ 8) generally happen on the water surface and long-chains (LCs) of PPFAS (C ≥ 8) generally emerge in sediments and the tissue of fish. They have an extensive array of undesirable impacts on HH, conditional on the exposure settings like the magnitude, exposure route, and time. Furthermore,

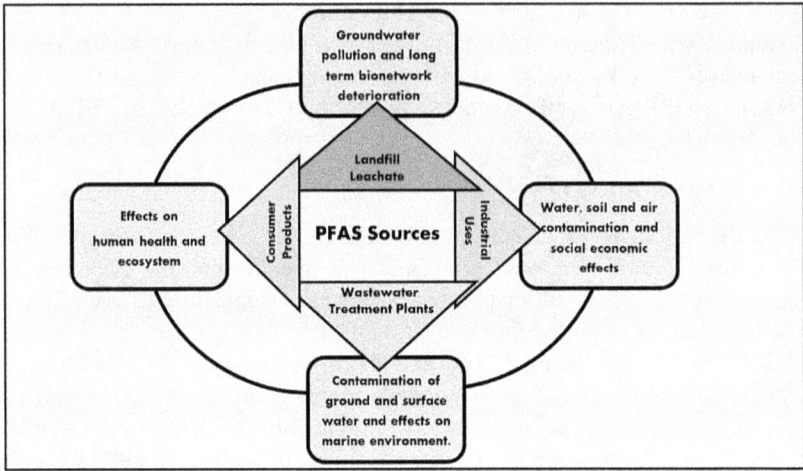

FIGURE 2.6   Key sources of PPFAS and their probable effects.

the physiognomies of the persons exposed also impact how severe the negative consequences could be (Habib, Song, Ikram, & Zahra, 2024).

PPFAS resident emissions and diffuse sources into the air are characterized by the packaging of food, textiles, medical equipment, building structures, paints, or the specialized utilization of FFs, manufacture of FPrs, printing paints, and inks. Furthermore, release stemming from the application and discarding of consumer products like personal care products, cosmetics, textiles, domestic products, food storage, and processing materials (Panieri, Baralic, Djukic-Cosic, Buha Djordjevic, & Saso, 2022).

Key causes of PPFAS contamination stem from the manufacturing and public-WWTPs, which openly contribute to the PPFAS discharged into the troposphere and brook system via the contaminated WW released or implicitly encourage PPFAS distribution into the soil via the diffusion of polluted sewage-sludge, reprocessed effluents and biosolids for agricultural applications. Other applicable sources of atmospheric discharge are signified by the manufacturing plants utilized for reprocessing and burning of PPFAS holding products or the landfilling of waste, which under explicit settings can percolate into soils, eventually inward bound into GW (Panieri, Baralic, Djukic-Cosic, Buha Djordjevic, & Saso, 2022). Furthermore, FF usage has been associated with PPFAS pollution in different ecological conditions like GW, soil, DW, sediment, biota, and SW under various practices and mechanisms. The historic WW release from firefighting drill events, lacking appropriate restraint, remediation, and disposal practices, has led to their straight disposal into the ecosystem (Perez, Lumpkin, Kornberg, & Schmidt, 2023; Teaf, Garber, Covert, & Tuovila, 2019).

Current information estimates that the number of possible holding installations that are discharging some amount of PPFAS is in the order of 1.0×10^5 or more in only the EU, a figure which might further be intensified in the succeeding years. Based on this information, it is not shocking that PPFAS have become prevalent ecological contaminants affecting the environment and the health of humans via the pollution of various pathways of human exposure, comprising of DW, cereals, fruits, atmosphere, milk, vegetables, and other food sources (Panieri, Baralic, Djukic-Cosic, Buha Djordjevic, & Saso, 2022). Figures 2.7 and 2.8 show the PPFAS sources

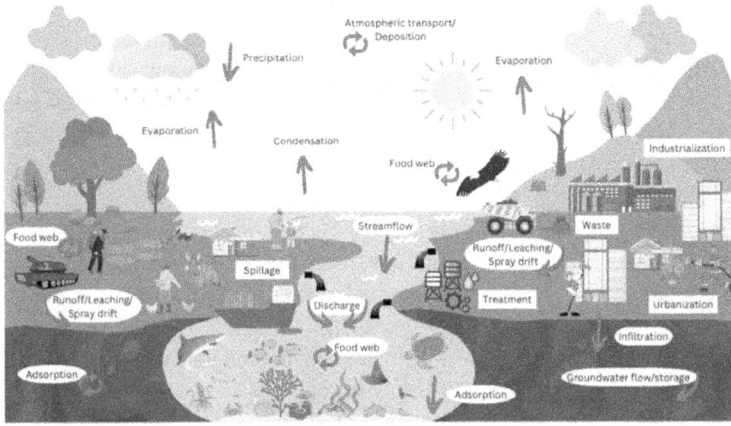

FIGURE 2.7 PPFAS sources and transport pathways in the environment. Adapted and reproduced from Perez et al. (2023) with permission from Elsevier B.V.

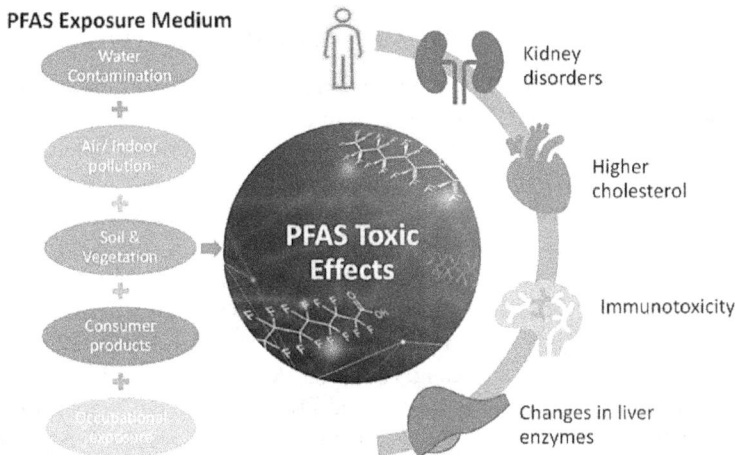

FIGURE 2.8 Outline of the media of PPFAS exposure and their noxious effects on humans. Adapted and reproduced from Habib et al. (2024) with permission from MDPI, Basel, Switzerland open access article distributed under the terms and conditions of the Creative Commons Attribution (CC BY) license.

and their transport routes in the ecosystem and the different PPFAS exposure media and their noxious effects on humans.

2.5 PPFAS ENVIRONMENTAL FATE

By virtue of the physicochemical properties and the extensive applications of PPFAS, they have been noticed in virtually all regions of the world, in various ecological media, human populations, and existing creatures. It is well-documented that PPFAS are considered to have by far the highest thermal-resistance, biotic, or chemical-degradation, which triggers ecological perseverance and poses grave apprehensions to the bionetwork and HH. Some polyfluorinated PPFAS can experience limited disintegration under certain ecological settings, and this can result in PPFAS substances developing with a superior effect than their precursors, as in the PFAAs case (Bilela et al., 2023; Habib, Song, Ikram, & Zahra, 2024).

A known illustration is characterized by the atmospheric oxidation of fluorotelomer alcohols, which leads to the creation of the consistent polyfluorinated aldehyde, which is further changed into PFCAs substances. It is further recognised that PFOAA and PFOSA can exist as impurities and can be directly released from manufacturing sources or consumer product discarding, therefore, derived from the degradation of biotic/abiotic or biotransformation processes of extended chain PPFAS molecules and other precursors like 8:2 FTOH. Furthermore, within the adipose tissue or the living organism's bloodstream, PPFAS are amphiphilic and can be bioaccumulated, and their elevated mobility renders their ecological distribution pervasive owing to leaching to GW, runoff into oceans and streams, wind distribution of dust particulates, and dry/wet deposition into soils (Habib, Song, Ikram, & Zahra, 2024).

2.5.1 EXPOSURE OF PPFAS TO HUMANS VIA DRINKING WATER (DW)

PPFAS pollution is extensive in both ground and SW, with levels characteristically extending from $ng.L^{-1}$ to $\mu g.L^{-1}$, as displayed in Figure 2.9. The main and recognized sources of PPFAS conveyance between the litho-sphere (soil) and the hydro-sphere (GW and SW) are military firefighting activities and facilities. PPFAS show a changing distribution in the bionetwork, impacted by their physicochemical properties, which are sequentially impacted by the C-chain length and function group (FG). Long-chain PPFAS generally show superior hydrophobicity and prove advanced adsorption to soil and sediment than petite-chain PPFAS do (Wee & Aris, 2023).

Human exposure to ecological toxicants happens generally via ingestion, dermal absorption, and inhalation (Xu, Kang, Gao, Lan, & Li, 2025). Exposure of humans to PPFASs happens via ingestion of polluted DW and seafood, indoor air inhalation, and interaction with other polluted media. They are regularly employed for their non-stick and surface-tension lowering features, which make them valuable for deterring oil and water (stain deterrence) and altering surface chemistry. The latter comprises of AFFFs, fluoropolymer production processing aids, semiconductor production, and metal plating applications. Figure 2.10 shows the outline of exposure

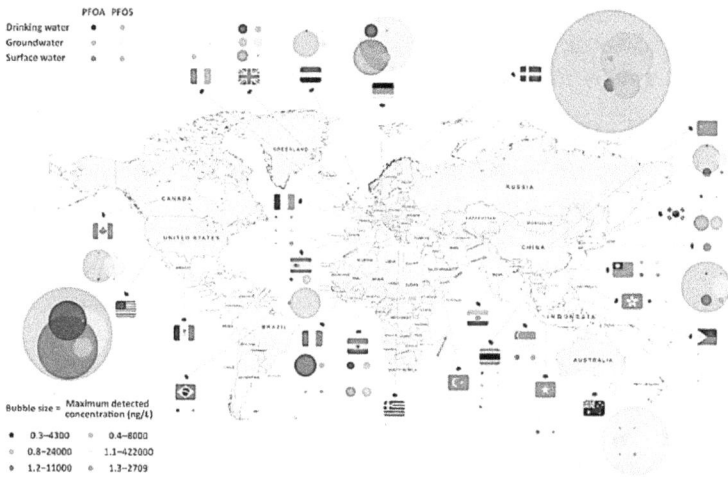

Fig. 3. Maximum concentrations of PFOA and PFOS detected in surface water, groundwater, and drinking water globally (see details in Supplementary Table C1). The occurrence and distribution of PFAS are influenced by historical and ongoing applications, as well as treatment efficiency, both of which are affected by various point and non-point stressors.

FIGURE 2.9 PPFOA and PPFOS optimum concentrations sensed in surface, ground, and DW, with the PPFAS occurrence and distribution being impacted by historic and current uses and treatment effectiveness, which are impacted by the different point and non-point sources. Adapted and reproduced from Habib et al. (2024) with permission from Elsevier B.V. open access article distributed under the terms and conditions of the Creative Commons Attribution (CC BY) license.

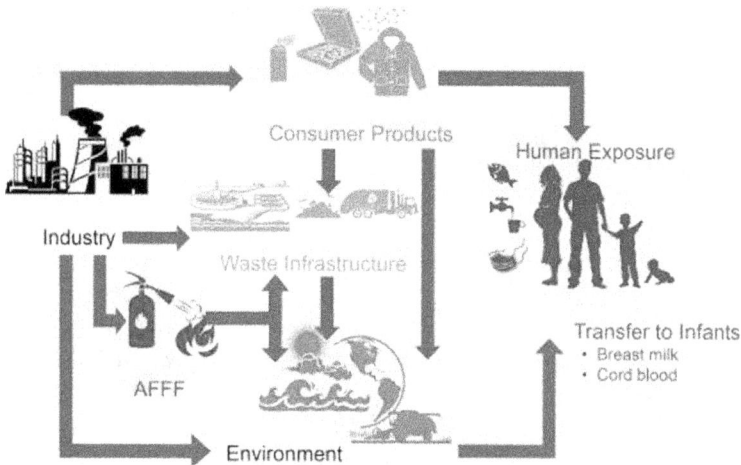

FIGURE 2.10 Outline of exposure pathways of PPFASs for various human populations, separate from work-related settings. Adapted and reproduced from Sunderland et al. (2019) with permission from Springer Nature.

pathways of PPFASs for various human populations separate from work-related settings (Sunderland et al., 2019).

The main way by which individuals are exposed to PPFAS is via their dietary consumption, followed by the ingestion of DW. GW, which is a key source of drinkable water in many regions, is substantially impacted by PPFAS pollution. This, in due course, impacts humans and animals upon ingesting contaminated water (Habib, Song, Ikram, & Zahra, 2024). The unswerving exposures owing to utilization in products can be swiftly phased out by changes in the manufacture of chemicals, but exposures motivated by PPFAS build up in the marine food chain and ocean, and the persistent pollution of GW with AFFF over extended periods. Considering the comparative standing of these various exposure routes is essential for considering drivers of temporal difference in serum PPFAS concentration assessed in biomonitoring research and for anticipating future exposure hazards (Sunderland et al., 2019).

A study of samples of DW from Burkina-Faso, Canada, France, Ivory-Coast, Chile, Mexico, the United States, Norway, and Japan showed three PPFAS (PFCAs, PFSAs, and perfluoroalkyl acid precursors) with concentrations extending from below recognition boundaries to 39 ng.L^{-1} (Kurwadkar et al., 2022). In this study by Sonnenberg et al. (2023), the drifts of serum PPFAS compounds such as PFOSA, PFOAA, PFHxSA, PFDA, and PFNAA were assessed to determine their dispersal among United States' populations. Based on the serum concentrations of PPFAS assessed from subjective subsamples of the National Health and Nutrition Examination Survey (NHANES) participants, it was found that adult males who were of Asian, non-Hispanic, Hispanic-blacks and -whites had an elevated risk of exposure to the designated PPFAS. It was also found in this study that from 1999–2018, the levels of serum PPFAS unceasingly decreased in the considered groups. Furthermore, it was observed that after 2016, the levels of the serum of PFOAA, PFDA, and PFHxSA exhibited an uphill drift in the smallest racial/ethnic group. Hence, the requirement for the continuous bio-observation of the level of PPFAS in the bionetwork and humans (Sonnenberg, Ojewole, Ojewole, Lucky, & Kusi, 2023). The association of PPFAS exposure with age-linked macular degeneration (AMD) was assessed in the study of Kang et al. (2024). It was stated in this study that 6.5% (132) of total participants were diagnosed as having any AMD, and this comprised of 5.7% (115) of participants with initial AMD. It was further observed that there was a substantial dose–response link noticed between the serum PFOSA concentration and any AMD for the high group when linked to the reference. Conclusively, the total PPFAS burden displayed a non-monotonic link with any AMD, and this might be an AMD risk factor in middle-aged and adults in the above-mentioned countries (Kang, Park, Kim, & Choi, 2024). Based on a study of PPFAS in commercial eggs from larger farmers across Denmark, it was found that there was no appearance or small PPFAS content in free-range and barn eggs (Granby, Ersbøll, Olesen, Christensen, & Sørensen, 2024). The target-screening (TS) and non-target-screening (NTS) were explored to assess human exposure to various chemical types in DW and urine samples. Based on the result of the analysis carried out utilizing LC-MS/MS, GC-MS/MS, and NTS with LC-HRMS, nine PPFAS were sensed in unfiltered tap water (TW) in the first sampling (median-concentration of 30-ng.L^{-1}) and six (median-concentration of 9.8 ng.L^{-1}) and five PPFAS in the second sampling and urine sample (Cserbik et al., 2023). In the study by Sáez et al. (2024), the existence of PPFAS in an urban aquifer in Barcelona (GW and river samples) and the processes that control their development along the GW flow were identified and assessed. Based on these samplings, the existence of 16 PPFAS products and three unique PPFAS were discovered. These short and extremely SC PPFAS were discovered to be pervasive, with the maximum concentrations observed for perfluorobutane sulfonic acid (PFBS), tri-fluoromethanesulfonic acid (TFSA) and tri-fluoroacetic acid (TFA). This long-chain PPFAS and unique PPFAS were discovered to exist in very extreme concentrations of < 50 ng.L^{-1} (Sáez et al., 2024). Sampled TW from the UK and China was assessed to determine the factors impacting PPFAS concentration in DW. It was found in this study that > 99% of PFOSA was regularly detected and dominated more in the global bottled water sampled, and 67–93% of other PPFAS were highly noticed. The PPFAS concentration in the sampled TW from China was 9.2-ng.L^{-1} when compared to the concentration of 2.7-ng.L^{-1} of sampled TW from the UK (Gao et al., 2024). In the study by Bao et al. (2019), elevated points of PPFAS pollution were reported in both SW and GW around a fluoro-chemical industrial park in Fuxin, China. The predominant PPFAS reported in this study were PFBS and PFOA in GW with optimum concentrations of 21.5- and 2.51-µg.L^{-1} (Bao et al., 2019). In the study of SW and sediments in Nevada, it was reported that the total PPFAS concentrations in the Truckee River and the Las Vegas Wash waters were 4.417×10^2 and 2.2343×10^3 ng.L^{-1}. The major species that were discovered in these waters were PFHxA (1.5–187.0-ng.L^{-1}), which was followed by perfluoro-pentanoic acid (PFPeA) (lower-detection-limit (LDL) to 169.9-ng.L^{-1}), PFOA (LDL to 65.5-ng.L^{-1}), and PFBS (LDL to 44.7 ng.L^{-1}). The overall PFAS in the Truckee River and Las Vegas Wash sediment were 272.9 and 345.7-µg.kg^{-1} (dry weight). The major species in the sediments were PFDS (LDL to 88.2-µg.kg^{-1}), PFHxA (LDL to 20.3-µg.kg^{-1}), PFBS (LDL to 29.1-µg.kg^{-1}) and perfluoro-undecanoic acid (PFUA) (LDL to 22.9-µg.kg^{-1}). SC PPFAS (C ≤ 8) and LC PPFAS (C > 8) were dominantly detected in water and sediment samples (Bai & Son, 2021).

2.5.2 PPFAS EXPOSURE TO HUMANS VIA SOIL AND VEGETATION

PPFAS contaminants enter the SW, GW, and soil environment from polluted sludge, industrial waste, or foaming agents that in the end enter the food chain via uptake by consumption of DW and plants (Ahmed et al., 2020). The adjacent watercourses also become contaminated by most PPFAS existing in manufacturing WW and landfill leachate (LL), in the end causing severe ecological hazards. The key inducement of PPFAS polluted GW or SW utilized for field irrigation are the application of biosolids as fertilizers in the agricultural sector. With the absorption of PPFAS by plants, petite-chain compounds are amassed in fruits and leaves and with extended chain compounds amassing in plant roots. Hence, they are integrated into the food webs (Habib, Song, Ikram, & Zahra, 2024). Based on the study of Wang et al., (2023), it was found that PPFAS in air and LLs were the main sources of PPFAS in soils. It was also found that PPFAS migrate from soil to plant via the uptake by roots, and this effect was found to vary with various PPFAS and species of plants. The highest concentration of PPFAS was noticed in the main exposure sites (10^{-1}-10^{2}-ng.g^{-1}), trailed by the secondary exposure sites and background sites with PPFAS concentrations of 10^{-2}-10^{1}-ng.g^{-1} and 10^{-2}-10^{1}-ng.g^{-1}, with the most predominant being legacy PPFAS-PFOAA with concentrations of 10^{0-1}-ng.g^{-1} and PFOSA 10^{0-2}-ng.g^{-1} (Wang, Munir, & Huang, 2023).

2.5.3 PPFAS EXPOSURE TO HUMANS DURING OCCUPATIONAL ACTIVITIES

A main concern of occupational exposure to PPFAS is the undesirable health impacts of PPFAS. High PPFAS serum concentrations are experienced by the workforce in specific professionals like firefighters, fluorochemical plant workers, and ski waxers. Owing to the exclusive PPFAS features that improve the product's quality and performance, polyfluorinatedalkyl compounds are uninterruptedly used in electronic products. But PPFAS application in electronic devices also increases the risk of PPFAS exposure via inhalation, ingestion, and skin absorption, for those managing and reprocessing electronic waste (e-waste). The work-related exposure pathways for PPFAS are the scavenging and recycling phases. Furthermore, the collection of e-waste and disposal and processing locations have been acknowledged as probable PPFAS exposure locations, as samples taken from these sites are PPFAS polluted (Habib, Song, Ikram, & Zahra, 2024).

In the study by Nilsson et al. (2022), present and former Australian firefighting services employees had their blood samples analyzed for PPFAS by means of HP LC-MS/MS. It was found that the mean serum concentrations of PFOSA were 27 ng.ml^{-1}, perfluoroheptane sulfonate (PFH$_p$S) was 1.7-ng.ml^{-1}, and perfluorohexane sulfonate (PFHxSA) was 14 ng.ml^{-1}, which were largely elevated than the level of PPFAS found in a wide-ranging Australian population (Nilsson et al., 2022). Based on the assessment of serum concentration for 14 PPFAS among Arizona healthcare emergency responders and other essential workers, it was found that firefighters had preeminent PFHxSA, branched and linear perfluorooctane sulfonic acid (PFOSA) and PFH$_p$S concentrations than other critical workers. It was also found that healthcare workers had higher odds of detection of branched (Sb) PFOAA and perfluorododecanoic acid (PFDoA) than other essential workers (Mitchell et al., 2025). In Liu et al. (2025), the identification of 64 PPFAS classes was assessed in serum samples from workers at a fluoro-chemical facility employing non-target approaches. Temporal drift analyses from 2008–2018 showed stable levels for most PPFAS but an upsurge in perfluorobutanoic acid (PFBA) and P-FHxSA was noticed. This was indicative of industrial shifts since the early 2010s from LC PPFAS to SC homologues in China (Liu et al., 2025).

2.5.4 PPFAS EXPOSURE TO HUMANS VIA CONSUMER-PRODUCTS, INDOOR-AIR, AND DUST

PPFAS in domestic substances contributes to human exposure when they migrate into dust, food, and indoor air. Carpets, paints, upholstery, outerwear, food, impregnation agents, cleaners, building materials, papers, polishes, and ski waxes have all been described to comprise of PPFAS. They can hypothetically leak into food from grease-impervious food wrapping, thereby snowballing dietary exposure. Also, various precursor chemicals exist in consumer products like cosmetics such as foundations, sunblocks, powders, etc. and can be biotransformed into PPFAS, and in the end, accrue in the body of humans. In the class of outdoor air contamination, production plants and work facilities are key sources of PPFAS discharges into the air, ensuing in the widespread pollution of surface and GW in the locality (Habib, Song, Ikram, & Zahra, 2024). Based on the assessment of 24 PPFAS and total F in dust from several rooms of 15 Massachusetts firefighting stations, it was found that turnout gear locker rooms had elevated dust levels of total F and three PPFAS (perfluorohexanoate [PFHxA], perfluoroheptanoate [PFHpA], and perfluorodecano-ate [PFDoDA]). These PPFAS were also discovered on six wipes of station turnout gear, and the foremost PPFAS in the living rooms was N-ethyl perfluorooctane sul-fonamide acetic acid (N-MeFOSAA), which is a precursor to PFOSA that persists, notwithstanding been phaseout, almost two decades ago (Young et al., 2021). In summary, PPFAS present a multifaceted hazard to ecological and community health, basically owing to its persistence, mobility, and bioaccumulative nature. Sustained monitoring, research, and worldwide cooperation are important to moderate the haz-ards linked with PPFAS.

2.6 CONCLUSION

PPFAS are a large class of artificial fluorinated compounds broadly utilized in manufacturing and consumer products owing to their thermal stability, chemical resistance, and surfactant properties. PPFAS are produced through electrochemi-cal ECF and TLN processes, leading to a comprehensive array of compounds like PFOAA and PFOSA. The main sources of PPFAS comprise of manufacturing facili-ties, WWTPs, landfills, and the FFs application, predominantly AFFFs. PPFAS has become a defining issue in ecological science and public health owing to its exten-sive application, chemical resilience, and mounting evidence of harm. Originally renowned for their exclusive properties like resistance to heat, oil, water, and chemi-cal degradation, they have found their way into immeasurable products and industrial processes over the past several decades. Nevertheless, this same resilience has led to a worldwide pollution catastrophe. PPFAS now persists in the air, water, soil, and even in human and wildlife bloodstreams. Their multifaceted chemistry and multi-plicity make them difficult to sense, regulate, and remediate. Moreover, production trends have shifted geographically, with many emerging countries now assuming the burden of industrial exposure, often with inadequate regulatory lapses. In conclu-sion, while the benefits of PPFAS in modern industry are indisputable, the ecological and health costs have become too important to ignore. Future efforts must balance technological progress with sustainable practices to protect bionetworks and com-munity health from the long-term consequences of these persistent contaminants.

REFERENCES

Ahmed, M., Johir, M., McLaughlan, R., Nguyen, L., Xu, B., & Nghiem, L. (2020). Per-and polyfluoroalkyl substances in soil and sediments: Occurrence, fate, remediation and future outlook. *Science of the Total Environment, 748*, 141251.

Ambaye, T., Vaccari, M., Prasad, S., & Rtimi, S. (2022). Recent progress and challenges on the removal of per-and poly-fluoroalkyl substances (PFAS) from contaminated soil and water. *Environmental Science and Pollution Research, 29*(39), 58405–58428.

Bai, X., & Son, Y. (2021). Perfluoroalkyl substances (PFAS) in surface water and sediments from two urban watersheds in Nevada, USA. *Science of the Total Environment, 751*, 141622.

Bao, J., Yu, W., Liu, Y., Wang, X., Jin, Y., & Dong, G. (2019). Perfluoroalkyl substances in groundwater and home-produced vegetables and eggs around a fluorochemical industrial park in China. *Ecotoxicology and Environmental Safety, 171*, 199–205.

Bilela, L., Matijošytė, I., Krutkevičius, J., Alexandrino, D., Safarik, I., Burlakovs, J., … Carvalho, M. (2023). Impact of per- and polyfluorinated alkyl substances (PFAS) on the marine environment: Raising awareness, challenges, legislation, and mitigation approaches under the One Health concept. *Marine Pollution Bulletin, 194*, 115309.

Brunn, H., Arnold, G., Körner, W., Rippen, G., Steinhäuser, K., & Valentin, I. (2023). PFAS: Forever chemicals—persistent, bioaccumulative and mobile. Reviewing the status and the need for their phase out and remediation of contaminated sites. *Environmental Sciences Europe, 35*(1), 1–50.

Buck, R., Franklin, J., Berger, U., Conder, J., Cousins, I., De Voogt, P., … van Leeuwen, S. (2011). Perfluoroalkyl and polyfluoroalkyl substances in the environment: Terminology, classification, and origins. *Integrated Environmental Assessment and Management, 7*(4), 513–541.

Buck, R., Korzeniowski, S., Laganis, E., & Adamsky, F. (2021). Identification and classification of commercially relevant per-and poly-fluoroalkyl substances (PFAS). *Integrated Environmental Assessment and Management, 17*(5), 1045–1055.

Cserbik, D., Redondo-Hasselerharm, P., Farré, M., Sanchís, J., Bartolomé, A., Paraian, A., … Flores, C. (2023). Human exposure to per- and polyfluoroalkyl substances and other emerging contaminants in drinking water. *npj Clean Water, 6*(1), 16.

Davidson, L., & Boland, M. (2023). Investigating three classification methods for per/polyfluoroalkyl substance (PFAS) exposure from electronic health records and potential for bias. *AMIA Summits on Translational Science Proceedings, 2023*, 52.

Dhore, R., & Murthy, G. (2021). Per/polyfluoroalkyl substances production, applications and environmental impacts. *Bioresource Technology, 341*, 125808.

Ehsan, M., Riza, M., Pervez, M., Li, C., Zorpas, A., & Naddeo, V. (2024). PFAS contamination in soil and sediment: Contribution of sources and environmental impacts on soil biota. *Case Studies in Chemical and Environmental Engineering, 9*, 100643.

Gaines, L. (2023). Historical and current usage of per-and polyfluoroalkyl substances (PFAS): A literature review. *American Journal of Industrial Medicine, 66*(5), 353–378.

Gao, C., Drage, D., Abdallah, M., Quan, F., Zhang, K., Hu, S., … Qiu, W. (2024). Factors influencing concentrations of PFAS in drinking water: Implications for human exposure. *ACS ES&T Water, 4*(11), 4881–4892.

Granby, K., Ersbøll, B., Olesen, P., Christensen, T., & Sørensen, S. (2024). Per-and polyfluoroalkyl substances in commercial organic eggs via fishmeal in feed. *Chemosphere, 346*, 140553.

Habib, Z., Song, M., Ikram, S., & Zahra, Z. (2024). Overview of per- and polyfluoroalkyl substances (PFAS), their applications, sources, and potential impacts on human health. *Pollutants, 4*(1), 136–152.

Kang, H., Park, S., Kim, D., & Choi, Y. (2024). Exposure to per- and polyfluoroalkyl substances and age-related macular degeneration in US middle-aged and older adults. *Chemosphere, 364*, 143167.

Kurwadkar, S., Dane, J., Kanel, S., Nadagouda, M., Cawdrey, R., Ambade, B., ... Wilkin, R. (2022). Per-and polyfluoroalkyl substances in water and wastewater: A critical review of their global occurrence and distribution. *Science of The Total Environment, 809*, 151003.

Liu, Y., Guo, Y., Lv, M., Wang, Y., Xiang, T., Sun, J., ... Liang, Y. (2025). Unraveling the exposure spectrum of PFAS in fluorochemical occupational workers: Structural diversity, temporal trends, and risk prioritization. *Environmental Science & Technology, 59*(12), 6247–6260.

Mitchell, C., Hollister, J., Fisher, J., Beitel, S., Ramadan, F., O'Leary, S., ... Ellingson, K. (2025). Differences in serum concentrations of per- and polyfluoroalkyl substances by occupation among firefighters, other first responders, healthcare workers, and other essential workers in Arizona, 2020–2023. *Journal of Exposure Science & Environmental Epidemiology, 2025*, 1–8.

Nilsson, S., Smurthwaite, K., Aylward, L., Kay, M., Toms, L., King, L., ... Bräunig, J. (2022). Serum concentration trends and apparent half-lives of per- and polyfluoroalkyl substances (PFAS) in Australian firefighters. *International Journal of Hygiene and Environmental Health, 246*, 114040.

Obodo, K. (2024). The application of magnetic sorbents for the sequestration of pesticides, pharmaceuticals, and perfluoroalkyl and polyfluoroalkyl substances from aqueous solutions. In: Ukhurebor, K.E., Aigbe, U.O. (eds) *Environmental Applications of Magnetic Sorbents* (pp. 6–1). IOP.

Panieri, E., Baralic, K., Djukic-Cosic, D., Buha Djordjevic, A., & Saso, L. (2022). PFAS molecules: A major concern for the human health and the environment. *Toxics, 10*(2), 44.

Perez, A., Lumpkin, M., Kornberg, T., & Schmidt, A. (2023). Critical endpoints of PFOA and PFOS exposure for regulatory risk assessment in drinking water: Parameter choices impacting estimates of safe exposure levels. *Regulatory Toxicology and Pharmacology, 138*, 105323.

Sabba, F., Kassar, C., Zeng, T., Mallick, S., Downing, L., & McNamara, P. (2025). PFAS in landfill leachate: Practical considerations for treatment and characterization. *Journal of Hazardous Materials, 481*, 136685.

Sáez, C., Bautista, A., Nikolenko, O., Scheiber, L., Llorca, M., Jurado, A., ... Pujades-Garnes, E. (2024). Occurrence and fate of perfluoroalkyl and polyfluoroalkyl substances (PFAS) in an urban aquifer located at the Besòs River Delta (Spain). *Environmental Pollution, 358*, 124468.

Schymanski, E., Zhang, J., Thiessen, P., Chirsir, P., Kondic, T., & Bolton, E. (2023). Per-and polyfluoroalkyl substances (PFAS) in PubChem: 7 million and growing. *Environmental Science & Technology, 57*(44), 16918–16928.

Sonnenberg, N., Ojewole, A., Ojewole, C., Lucky, O., & Kusi, J. (2023). Trends in serum per- and polyfluoroalkyl substance (PFAS) concentrations in teenagers and adults, 1999–2018 NHANES. *International Journal of Environmental Research and Public Health, 20*(21), 6984.

Su, A., & Rajan, K. (2021). A database framework for rapid screening of structure-function relationships in PFAS chemistry. *Scientific Data, 8*(1), 14.

Sunderland, E., Hu, X., Dassuncao, C., Tokranov, A., Wagner, C., & Allen, J. (2019). A review of the pathways of human exposure to poly-and perfluoroalkyl substances (PFASs) and present understanding of health effects. *Journal of Exposure Science & Environmental Epidemiology, 29*(2), 131–147.

Tang, Z., Hamid, F., Yusoff, I., & Chan, V. (2023). A review of PFAS research in Asia and occurrence of PFOA and PFOS in groundwater, surface water and coastal water in Asia. *Groundwater for Sustainable Development, 22*, 100947.

Teaf, C., Garber, M., Covert, D., & Tuovila, B. (2019). Perfluorooctanoic acid (PFOA): Environmental sources, chemistry, toxicology, and potential risks. *Soil and Sediment Contamination: An International Journal, 28*(3), 258–273.

Wang, Y., Munir, U., & Huang, Q. (2023). Occurrence of per- and polyfluoroalkyl substances (PFAS) in soil: Sources, fate, and remediation. *Soil & Environmental Health, 1*(1), 100004.

Wang, Z., Buser, A., Cousins, I., Demattio, S., Drost, W., Johansson, O., … White, G. (2021). A new OECD definition for per- and polyfluoroalkyl substances. *Environmental Science & Technology, 55*(23), 15575–15578.

Wang, Z., DeWitt, J., Higgins, C., & Cousins, I. (2017). A never-ending story of per- and poly-fluoroalkyl substances (PFASs)? *Environmental Science & Technology, 51*, 2508–2518.

Wee, S., & Aris, A. (2023). Environmental impacts, exposure pathways, and health effects of PFOA and PFOS. *Ecotoxicology and Environmental Safety, 267*, 115663.

Xu, B., Liu, S., Zhou, J., Zheng, C., Weifeng, J., Chen, B., … Qiu, W. (2021). PFAS and their substitutes in groundwater: Occurrence, transformation and remediation. *Journal of Hazardous Materials, 412*, 125159.

Xu, H., Kang, J., Gao, X., Lan, Y., & Li, M. (2025). Towards a better understanding of the human health risk of per- and polyfluoroalkyl substances using organoid models. *Bioengineering, 12*(4), 393.

Young, A., Sparer-Fine, E., Pickard, H., Sunderland, E., Peaslee, G., & Allen, J. (2021). Per- and polyfluoroalkyl substances (PFAS) and total fluorine in fire station dust. *Journal of Exposure Science & Environmental Epidemiology, 31*(5), 930–942.

3 Environmental Fate and Mobility of Perfluoroalkyl and Polyfluoroalkyl Substances

3.1 INTRODUCTION

The most common per- and polyfluoroalkyl substances (PFAS) discovered in the environment include a compound comprising of a polar functional group (FG) and a carbon-fluorine tail. The compound tail is hydrophobic and often lipophobic, whereas the head groups may be polar and hydrophilic (HDP) (Redmon et al., 2025; Glüge et al., 2020; Dawson et al., 2023; Buck et al., 2011). The opposing inclinations of the head, as well as the tail, may cause the surroundings to be widely yet unevenly distributed. Multiple PTN processes should be taken into account when describing the fate and transit of PFAS due to the complexity of subterranean environments, soils with various surface charges, organic carbon (OC), interfaces between air and water, and interfaces with water and hydrocarbon (HC) co-contaminants. Additionally, depending on the concentration (CCN), PFAS may behave differently. For example, they may have a propensity to form MCs in high CCNs. Even though PFAS's structure usually renders them resistant to water and oil in many goods (such as dry surface coatings), their propensity to partition to phospholipid bilayers (bacterial membranes) suggests that they may not have lipophobic tendencies in the aqueous phase (Fitzgerald et al., 2018; Jing et al., 2009).

Interfacial behaviours, hydrophobic effects, and electrostatic interactions are significant PFAS PTN processes. One of the main mechanisms influencing the interaction with OC in soils is electrostatic effects. For example, the net negative surface charge of natural soils and aquifer materials typically attracts cationic or zwitterionic PFAS while rejecting the perfluoroalkyl benzamides (PFAAs) negatively charged heads, which are frequently existent in environmental media as anions. Like any surfactant, the head and tail groups' competitive natures cause a buildup along environmental media interfaces (Brusseau, 2018; Guelfo & Higgins, 2013; McKenzie et al., 2016).

DOI: 10.1201/9781003625537-3

As rightly reported by Garg et al. (2023), there are several possible environmental and health impacts of the exposure of PFAS (whose source is mainly from the industries) to both humans and animals, as illustrated in Figure 3.1. Also, they further reported that as a result of the inappropriate management and dumping of waste, which can in turn adulterate both ground and surface water, the exposure path for PFAS in the environment is shown in Figure 3.2. However, more details about several possible environmental and health impacts of exposure to PFAS, as well as the exposure path for PFAS in the environment, the general analysis scheme and procedures for PFAS in the environment (Figure 3.3), and the possible ways to prevent the exposure to PFAS in the environment (Figure 3.4) are contained in Garg et al. (2023). Hence, this chapter discusses the environmental fate and mobility of PFAS in the environment.

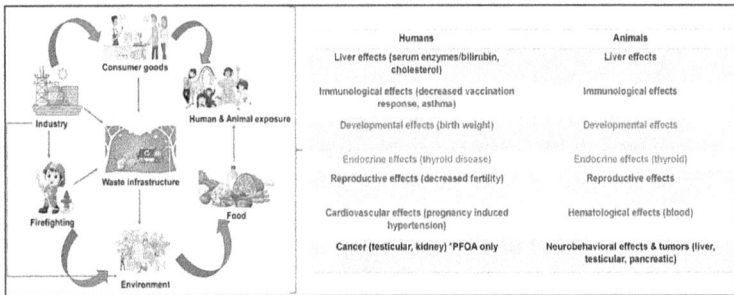

FIGURE 3.1 Possible environmental and health impacts of the exposure of PFAS to both humans and animals. Adapted and reproduced from Garg et al. (2023) with permission from Elsevier B.V.

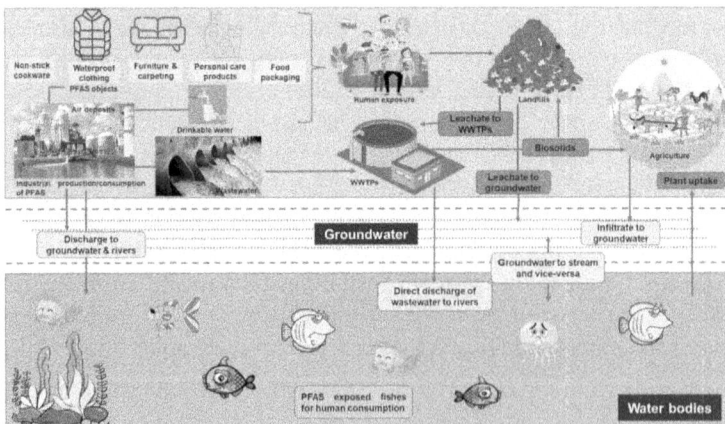

FIGURE 3.2 The exposure path for PFAS in the environment. Adapted and reproduced from Garg et al. (2023) with permission from Elsevier B.V.

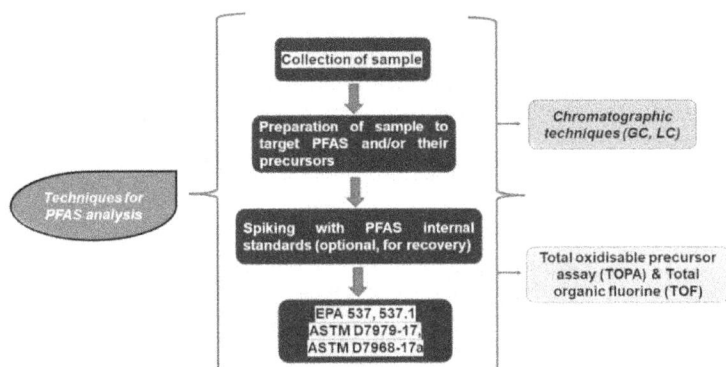

FIGURE 3.3 The general analysis scheme and procedures for PFAS in the environment. Adapted and reproduced from Garg et al. (2023) with permission from Elsevier B.V.

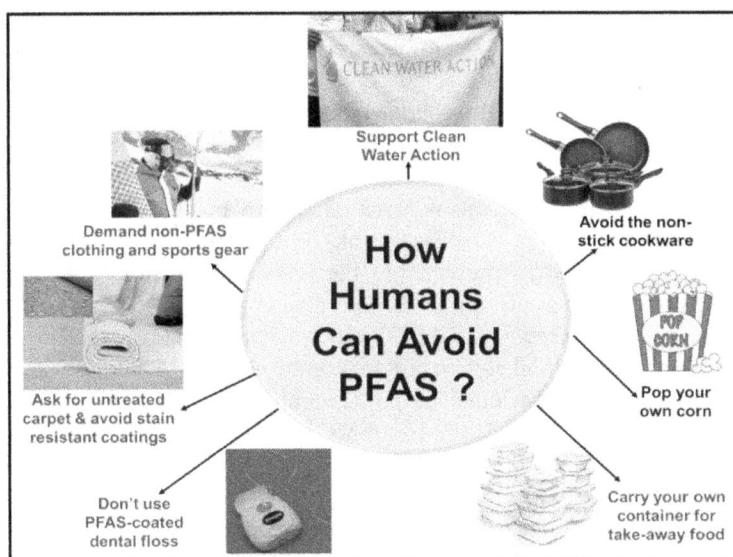

FIGURE 3.4 The possible ways to prevent exposure to PFAS in the environment. Adapted and reproduced from Garg et al. (2023) with permission from Elsevier B.V.

3.2 CONSIDERATIONS FOR PFAS PTN

3.2.1 PURE PHASE PFAS

It is rare for PFAS to arise in the environment as a distinct phase (such as solid PFAS, light non-aqueous phase liquids (NAPL) PFAS, or dark NAPL PFAS) because of its excellent aqueous solubility. While solid salts containing PFAS can exist, most product applications utilize miscible solutions, often combinations of several chemicals.

Several of these substances have comparatively high-water solubility (Ney, 1995). For instance, according to the USEPA, (2017d), PFOA has a solubility of 9.50×10^3 mg/L at 25.00 °C. It should be noted that there have been reports of interactions between PFAS and NAPL co-contaminants in the subsurface, which may affect migration there.

3.2.2 PFAS FORMATIONS OF MCs AND FOAM

A surfactant is a material that, when dissolved in a liquid, tends to reduce the liquid's surface tension. Some PFAS compounds have the ability to function as surfactants, which can cause PFAS molecules to group together to form supramolecular aggregations. Additionally, the reason for accumulation at interfaces, such as the air-water contact, is the surfactant characteristics of PFAS. The heat resistance and surfactant qualities have been used for applications like AFFF, which covers the fuel source with a thin layer of water.

According to experimental data, PFAS can function as ionic HC surfactants in supramolecular aggregations, forming MCs at a critical MC CCN (CMC) (Downer et al., 1999; Pedone et al., 1997). However, other theories indicate that, in the case of the production of ionic HC surfactants and hemi-MCs might start at CCNs as low as 1.00×10^{-3} times the CMC (Kancharla et al., 2022; Yu et al., 2009; Johnson et al., 2007). Additional investigation is required about the propensity of PFAS to form conventional MCs, such as oil-in-water emulsions, as specific results indicate that the behaviour of ionic HC surfactants may not be directly comparable to that of PFAS supramolecular aggregation (Costanza et al., 2019). Research has only started to address the extra complexity presented by the associated functions of various environmental factors, such as pH, ionic CCN, co-contaminant presence, etc.

With the possible exception of AFFF source releases, the recognized CMCs of PFAAs are significantly higher than ordinary ambient CCNs, leading some research to conclude that PFAA behaviour at the CMC is unlikely to have any practical impact (Brusseau, 2018; Horst et al., 2018). Hemi-MC production of PFAAs, however, can start at CCNs as low as 1.00×10^{-3} times the CMC, much like with ionic HC surfactants (Yu et al., 2009; Johnson et al., 2007). This may be important to comprehending PFAA sorption for developing possible treatment methods, environmental destiny, and transport. The degree of sorption can be influenced by the interaction of "supramolecular aggregations" of PFAAs with both hydrophobic and charged surfaces, however, other parameters determine the extent and even the direction of effect (Deng et al., 2012; Du et al., 2014; Zhang et al., 2019). Adsorption (AST) of the HDP regions of PFAAs, known as the "heads", onto "positively charged absorbent surfaces", for example, may aid in creating and aggregating hemi-MCs on surfaces. Furthermore, plugging the intraparticle pores of porous surfaces with hemi- or even MCs may reduce the quantity of sorption.

Surface water can accumulate PFAS in its surface microlayer (SML), a thin layer of water (50 μm) in contact with the surrounding air. This accumulation may cause foam to form above the water's surface due to waves, winds, or other turbulent forces carrying air. Even at PFAS, CCNs are lower than recorded CMCs (Schwichtenberg et al., 2020). This foam differs from AFFF since it most possibly results from surface water body agitation and the aggregation of PFAS in the dissolved phase. This kind of foam above the water's surface that contains PFAS has been seen to occur close to or downstream of locations where PFAS have been spilled (AECOM, 2021).

3.2.3 PTN to Solid Phases

Soils and sediments as reported by Higgins & Luthy (2006), sewage solids as reported by Ebrahimi et al. (2021), biosolids as reported by Venkatesan & Halden (2013), organic matter as reported by Fitzgerald et al. (2018), and iron oxides in acidic ecosystems as reported by Campos-Pereira et al. (2020), are among the solid-phase materials to which PFAS can partition. PFAS are retained on the solids by PTN, which slows PFAS migration in the saturated zone and lessens or delays PFAS leaching from sources in the vadose zone. On the other hand, PFAS may split into mobile colloids, which would facilitate transit instead of slowing it down (Brusseau et al., 2019). In circumstances where colloidal conveyance is more likely to ensue, for instance, sediment deposition as well as movement in flowing water and wet–dry or freeze–thaw cycle settings, this may be particularly important (Borthakur et al., 2021).

3.2.4 PTN Processes

There has been much research on the PFAS PTN to soils and solids, and two main mechanisms are believed to be involved: the two methods of sorption: 1) electrostatic interactions and 2) hydrophobic interactions for sorption to organic materials (Higgins & Luthy, 2006). The comparative contribution of each procedure may differ depending on the level of the PFAS, the composition of the soil, the surface chemistry, ion CCN, as well as other geochemical characteristics. PFAS PTN to solids might be unexpected and vary depending on site-specific conditions due to the wide range of PFAS compositions and carbon chain lengths.

3.2.5 Electrostatic Interactions

The kind of soil and its chemistry—specifically, pH and the presence of polyvalent cations—substantially influence the PTN process via electrostatic interactions. Most soils include both fixed-charge and variable-charge surfaces, which implies that pH may have a considerable influence on the soil's net charge and the charge of certain PFAS FGs. For example, the cationic FGs on certain PFAS may interact electrostatically with the net negative charge of the majority of clay particles (Barzen-Hanson et al., 2017).

By changing charges on the surface or even the ionic makeup of PFAS, pH variations may have an impact on these electrostatic processes (Nguyen et al., 2020). Anionic PFAS, such as PFOS as well as other PFAAs, have been shown to sorb more readily in soil solutions with higher calcium CCNs and lower pH levels (Higgins & Luthy, 2006), even if some soils' capacity to function as a buffer, such as those containing carbonate minerals, may mitigate pH changes. Increased PTN of certain PFAS to soil can also result from higher CCNs of certain polyvalent cations, such as Ca^{2+}, Mg^{2+}, and Fe^{2+}, which sorb strongly to permanent charge sites on clay minerals (Higgins & Luthy, 2006; McKenzie et al., 2015). Charge variations in the PFAS FGs have been linked to the considerable variation in the effect on cationic and zwitterionic PFAS (Mejia-Avendaño et al., 2020). Iron-oxide has been linked, in the literature, to anionic PFC sorption; nonetheless, the primary factor affecting sorption is the organic matter content (Higgins & Luthy, 2006).

3.3 SORPTION TO ORGANIC MATTER

In Higgins and Luthy (2006), in these and previous investigations, OC was discovered to be the primary factor influencing the degree and kinetics (KNT) of PFAS sorption, with OC ranging from 0.56% to 9.66% (Higgins & Luthy, 2006; IARC, 2016; Li et al., 2018; Sima & Jaffé, 2021; Wei et al., 2017). Both PFCAs and PFSAs often bind to the portion of sediment or soil that is made up of OC (Guelfo & Higgins, 2013; Higgins & Luthy, 2006) despite being anions at ambient pH levels, they are very mobile in saturated and unsaturated zones of groundwater (Xiao et al., 2015).

As perfluoroalkyl tail length rises, so does the overall sorption to OC, which is correlated with increased hydrophobicity (Cai et al., 2022; Guelfo & Higgins, 2013; Higgins & Luthy, 2006; Pereira et al., 2018; Sepulvado et al., 2011), showing that PFSAs (like PFBS) and PFCAs (like PFBA) with short chains are less retarded than those with long chains (PFOS and PFOA, respectively). PFSAs sorb more strappingly than PFCAs with comparable chain lengths, in addition to the linear isomers being more sorptive than branching isomers (Higgins and Luthy 2006; Karrman et al. 2011).

3.3.1 PTN COEFFICIENTS

The range of OC PTN coefficients (Koc) for persistent flame retardants (PFAS) that can be found in the environment. Koc is a normalized AST coefficient for soil OC and might be a helpful metric to assess possible retardation. Nevertheless, no direct contributions from electrostatic interactions are captured by this parameter (Higgins & Luthy, 2006), this means that, depending on the PFAS under consideration, the estimation of sorption via the measurement of the fraction of OC (foc) in the soil in conjunction with a Koc value taken from the literature may underestimate retardation. Additional site-specific data, including retardation coefficients from electrostatic processes, pH, and the presence of polyvalent cations are also considered.

Li et al. (2018) analyzed data from several studies. They determined that the majority of PTN coefficients (Kd) calculated for different PFAS were most closely connected to OC content and pH. Nevertheless, Barzen-Hanson et al. (2017) in their study revealed that there is a general deficiency of relationship between the partition coefficients determined for several PFAS, including many cationic and zwitterionic PFAS and anionic fluorotelomer sulfonates, and soil characteristics (such as

OC). Anderson et al. (2019) examined field data from several AFFF release locations and concluded that OC significantly affects the ratios of PFAS soil-to-groundwater CCNs. Based on these findings, they utilized statistical modelling to determine apparent Koc values for 18 distinct PFAS.

Since steady state conditions are seldom seen in real systems, using PTN coefficients to determine sorption necessitates making these assumptions. In idealized systems, equilibrium is frequently established over several days to weeks, however, PFAS sorption KNT varies by FG, length of the carbon chain, and soil type (Schaefer et al., 2021; Xiao et al., 2017a). Studies conducted in labs and through models have demonstrated that in some situations, rate-limited sorption factors make it difficult to simulate the PTN of PFAS using equilibrium sorption parameters (Brusseau, 2020; Guelfo et al., 2020). For instance, by overestimating the amount of desorption or underestimating the retardation factor, lab-based Kd values may understate the influence of sorption during fate and conveyance modelling. According to Schaefer et al. (2021), the quantity of desorption from soils exposed to PFAS years ago was often far below what the reported Koc foc equations for the PFAS in question suggested. This indicates that desorption from prior PFAS exposures could possibly have fewer noticeable effects on groundwater, especially for shorter chain PFAS. A significant percentage of the PFAS in the historically polluted soils by AFFF in the research were not readily leachable utilizing the selected desorption techniques, according to Schaefer et al. (2022).

3.3.2 Nonlinear Sorption, Hysteresis, and Mass Transfer Limitations

Hysteresis, mass-transfer constraints, and non-linear sorption can all affect PFAS sorption processes. According to nonlinear sorption, PFAS would normally sorb more strappingly at small PFAS CCNs than high CCNs. At least one field investigation found that sorption increased with increasing CCNs, in contrast to usual nonlinear sorption tendencies (Anderson et al., 2022). Additionally, there is an indication that desorption for certain PFAS could possibly happen more sluggishly than sorption; this is thought to be the consequence of diffusion limits and trapping (Chen et al., 2016a; Higgins & Luthy, 2006; Xiao et al., 2019; Zhi & Liu, 2018). According to several studies, certain PFAS may be prone to rate-limiting sorption (such as diffusion restricted), which means that some of the PFAS would possibly desorb more sluggishly than others (Schaefer et al., 2021; F. Xiao et al., 2019). As per the findings of (Brusseau et al., 2019a), PFOS demonstrated non-ideal sorption/desorption behaviour, namely tailing. According to the study's KNT desorption model, "the rate of desorption was proportional to the PFAS aqueous diffusivity" (Schaefer et al., 2021), corroborating the notion that the rate of release from soil may be constrained by diffusion. It is anticipated to be less bioavailable and migratory, but it may be more persistent if released PFAS is strappingly reserved in sediments or the soil matrix. According to the explanation for certain PFAS's natural attenuation, which was examined by Newell et al. (2021a), this hysteresis effect on the percentage of PFAS that has been absorbed in soils may also have aided in sorption.

- In certain cases, it has been demonstrated that PFAS PTN due to electrostatic interactions is non-linear; Xiao et al. (2019) established that altering the coefficient of sorption for a number of PFAS to soils suggests a decrease in PFAS CCNs. The notable non-linearity of PFAS in zwitterionic systems noted by Xiao et al. (2019) showed a significant hysteresis as well, which was assumed to be caused by entrapment unrelated to surface complexation or soil organic matter. The significance of electrostatic interactions with minerals for PFAS, for example, PFOA and PFOS, which are anionic substances with ecologically significant pH values, can be determined via the measurement of the anion exchange ability in typical soils. However, given the wide range of pH values, pH may also be a useful indicator of PFAS mobility in soil. Measuring the ionic strength or cation CCN of aqueous solutions may also help elucidate the potential role of electrostatic attraction in increasing PTN.

The results covered in this subsection are especially pertinent to the powerfully sorbing long-chain PFAS, such as PFOS (Chen et al., 2016b) and PFAS that have been absorbed and show significant hysteresis, like zwitterionic PFAS. These PFAS have significant implications on the fate and conveyance of PFAS, including leaching from soil to groundwater, movement as well as deceleration in the saturated region, and if sorption could operate as a natural reduction mechanism (Newell et al., 2021a, 2021b). To determine if any of these bulk factors may be utilized predictably for destiny and transport investigations, more investigation is required (Newell et al., 2021a) and to comprehend the part that the factors play in the irreversible sorption of decades' worth of PFAS. The inference that basic correlations of sorption with OC or pH are not enough to envisage PFAS PTN coefficients for a broad range of PFAS is supported by current research. Site-specific data could possibly be more pertinent for comprehending PFAS conveyance in the absence of a consistent model that forecasts the influence of diverse processes on the degree of sorption (and hysteresis) (Anderson et al., 2016; Knight et al., 2019; Li et al., 2018), as well as considerations for pump-and-treat systems.

3.3.3 PTN to Air

At the moment, there is a dearth of trustworthy information on physical characteristics linked to PFAS volatilization, namely vapour pressure and Henry's law constants (Kaw). The collection is expanding, though. Moreover, acidic PFAS, including PFAAs, volatilize from water to air due to pH-dependent aqueous phase dissociation from more volatile acidic to less volatile anionic species (Kaiser et al., 2010). The acidic forms of perfluorononanoic acid (PFNA), PFOA, perfluorodecanoic acid (PFDA), perfluorododecanoic acid (PFDoDA), perfluoroundecanoic acid (PFUnA), and FTOHs are among the specific PFAAs for which measured vapour pressures are known (Kaiser et al., 2005; Barton et al., 2008). For a particular PFAS, Kaw can vary across many orders of magnitude; the underlying reasons for the wide range of stated values are unclear, although they may have to do with variable testing settings. The reported vapour pressures and water solubility of PFAS, such as PFOS and PFOA, are low and high, respectively, which restricts the volatilization of water to air. These PFAS are acidic at pH values that are significant to the environment (USEPA, 2000). Roth et al. (2020a) found that some FGs in other PFAS, such as FTOHs, confer higher volatility. It has been shown that stirring AFFF releases gas-phase PFAS, which allegedly contains PFOA. Five FTOHs and ten PFCAs were found to have detectable amounts above background laboratory air values in these controlled lab trials. The validity of the PFOA detections has been disputed, and it was not determined if this kind of PTN could happen in field conditions (Roth et al., 2020b; Titaley et al., 2020).

Under some circumstances, PFAS could possibly be released into the atmosphere and spread throughout it, especially when it comes to emissions from industrial stacks and when fires are suppressed, or burned. Anionic PFAS that may be absorbed by particles (Ahrens et al., 2012) and perhaps gas-phase volatiles such as FTOHs (Thackray et al., 2020) may be present. In the latter case, anionic PFAS interact with airborne aerosols as well as other minute particles to facilitate movement instead of immediately PTN to the gas stage. In semi-rural and urban areas, for example, airborne particulate matter (PM) has been reported to include both PFOA and PFOS, with PFOA being more prevalent in the larger, coarser fractions and PFOS in the smaller, ultrafine particles (Ge et al., 2017). According to these investigations, certain PFAS can adsorb to particles, probably due to the effect of diffuse local sources. Wet and dry deposition techniques that scavenge gaseous PFAS that have partitioned into water droplets or particle-bound PFAS are two ways to remove PFAS from the air (Dreyer et al., 2010). When PFAS are washed away by raindrops and fall to the ground without being transferred to the atmosphere, this phenomenon is known as wet deposition (Barton et al., 2007). Within a few kilometres downstream of a major production source, this process has been shown to be essential for influencing airborne transport (Barton et al., 2007). Dry deposition is a naturally occurring process that relies on particle characteristics and environmental conditions. Since these deposition processes eliminate PFAS from the atmospheric environments, they could have an impact on the amount and location of PFAS deposition in both terrestrial and aquatic environments.

The PTN of PFOA from surfaces that wet and dry into the air at work was demonstrated by Kaiser et al. (2010). These substrates' PTN to air is influenced by internal conditions; lower pH ecosystems contribute more PFOA to the air. The increased vapour pressure of PFOA in its protonated acid form may account for these findings (Kaiser et al., 2005). It is interesting that these researchers proved that PFOA separates from surfaces that are dry more easily than from water, which might lead to a considerable increase in occupational exposure.

3.3.4   PTN to Air/Water Interfaces

As previously mentioned, PFAS frequently behave as surfactants since they have both hydrophobic and HDP characteristics. These characteristics have complicated effects on transportation, which are currently being researched. Many PFAS, if present at higher quantities, will diminish interfacial tension and specifically form films at the air-water interface, with the HDP head group dissolved in the water and the hydrophobic carbon–fluorine (C–F) tail directed in the direction of the air (Krafft & Riess, 2015). This characteristic affects the transit and deposition of aerosols, indicating that PFAS will accumulate on the surfaces of water bodies (Prevedouros et al., 2006).

The vadose zone, where unsaturated settings deliver a large air-water interfacial space, is where this preference for the air-water interface has the biggest impact on PFAS transmission (Brusseau & Guo, 2022; Brusseau et al., 2019a; Brusseau, 2018). This includes the potential for improved retention at the capillary fringe and vadose zone, which are the subject of several recent research in the aspects . For instance, Brusseau (2018) found that PFOS and PFOA AST at the air-water interface might raise the retardation factor for aqueous-phase movement, resulting in more than half of the entire retention in a model setting.

Air-water PTN may, therefore, aid in reducing the pace at which PFAS spreads in unsaturated soil. According to Anderson et al. (2019), soils with greater clay levels had lower ratios of soil-to-water CCNs for specific PFAS. The authors reasoned that the increased water content inside these clay-rich zones would offer a more credible explanation than other depth-discrete zones with coarser-grained material. As a result, less space would be available for PFAS PTN between air and water, which would lower soil retention in general. The hypothesis that negatively charged surfaces of clay might lessen anionic PFAS AST via electrostatic repulsion is also supported by this pattern. Guo et al. (2020) emphasized the significance of the kind of soil on the air-water interfacial building as a retention procedure in a modelling study, which found that, in their experimental settings, retardation factors for PFOS ranged from 146 to 792 for finer-grained soils and from 233 to 1,355 for sands. This was ascribed to the decreased capillary interactions/forces in the sands, which led to lower water levels as well as a better air-filled pore space to help with PFAS PTN. Because the study solely examined PFOS and did not use input parameters from the field, it should be noted that the simulations conducted by Guo et al. (2020) should be utilized for extending these findings to the more well-known family of PFAS. Aside from the soil type effect noted previously, ionic intensity has also been found to alter air-water interfacial PTN; in unsaturated soils, PFAS deceleration seems to be exacerbated as salinity increases (Costanza et al., 2020; Le et al., 2021; Lyu & Brusseau, 2020).

Interest in including these processes in predicted destiny, transport models, and CSMS has increased. A quantitative foundation for these kinds of models can be supplied by direct measurements of mass discharge to groundwater, estimates of the constituent-specific air-water AST coefficients and air-water interfacial area (Brusseau, 2019a,b; Guo et al., 2020). When a PFAS CCN rises, its air-water AST coefficients often decrease, and the process seems non-linear (e.g., Freundlich-type PTN) (Schaefer et al., 2019). The quantity of perfluorinated carbons also tends to raise air-water AST coefficients (Brusseau, 2019b; Schaefer et al., 2019). The AST of the other, a reduced amount of surface-active PFAS in the mixture is anticipated to be decreased by the preferential AST of the more surface-active PFAS in a mixed release system, which might affect the relative breakthrough durations of various PFAS (Silva et al., 2021). According to Silva et al. (2021), this suggests that when PFAS are released as a multicomponent mixture, these procedures would have varying effects on each molecule, with shorter-chained PFAS (such as PFBS and PFPeA) being maintained at lower levels than longer-chained PFAS (such as PFOS, PFDA, and PFNA).

The potential for PFAS retention in the vadose zone as a result of AST at the air-water interface is a significant factor that should be covered in the CSM, despite the fact that PFAS retention varies by site. In soils with a usually low water content that is in the vadose zone or shallow groundwater, the importance of air-water interfacial PTN may be lessened. To understand how these mechanisms impact PFAS migration across the vadose region at a specific site, in-depth site investigations are necessary. For example, a thorough examination of the levels of PFAS in the soil at contaminated sites showed that, while the vertical movement of PFAS to the water table persisted, PFAS were considerably maintained in the vadose zone for prolonged periods (Brusseau, 2020).

3.3.5 PTN into NAPL Co-Contaminants

Nonaqueous Phase Liquids

Petroleum (PTL) HC fuels in the form of NAPLs and PFAS could mix on fire training grounds, fire response sites, as well as other places where fuels have been utilized or disposed of alongside PFAS-comprising products. In these conditions, the discharged PTL HC fuel forms a NAPL, from which PFAS can separate and collect at the NAPL/water interface, as reported by Brusseau (2018). These procedures might lead to better PFAS persistence, higher PFAS sorption onto the NAPL/water interface and ensuing retardation, and enhanced PFAS mass retention in NAPL source zones (Brusseau, 2018; McKenzie et al., 2016). Depending on the specifics of each site, this process may contribute differently to other PFAS PTN mechanisms (such as the solid-stage and air-water interface). Nevertheless, some studies have demonstrated that the PFAS mass accruing at the NAPL-water interface would possibly be lower than that at the air-water interface in systems that contain all four stages (Brusseau, 2019a; Costanza et al., 2020; Silva et al., 2019).

There may be further PFAS consequences from NAPL. The oxidation-reduction and biogeochemical conditions under the surface can be dramatically changed by PTL light NAPL, which are biodegradable NAPL. Subsurface PTL light NAPL from a PTL-based fire, for instance, may increase methane CCNs while lowering those of oxygen as well as other electron acceptors in the vicinity. Anaerobic transformation activities may occur in the confined area of anoxic reduction circumstances that the light NAPL produces, inhibiting the aerobic transformation processes of PFAS.

3.4 MEDIA-SPECIFIC MIGRATION PROCESSES

This topic includes a discussion of the potential impacts of air movement, soil leaching into groundwater, and diffusion into matrices with limited permeability. The

following processes occur inside a specific medium, as opposed to the PTN procedures that encompass the transport of chemicals between media. When it comes to
PFAS migration, these can be important considerations.

3.4.1   Diffusion In and Out of Lower Permeability Materials

Molecules move in response to a CCN gradient through a process called diffusion.
Since diffusion rates in groundwater are slower than those in advection, diffusion is
frequently disregarded. Molecular diffusion is the primary mechanism behind the
migration of pollutants in low-permeability materials, and depending on the strength
of the hydraulic gradient, advective effects may not be as significant. Analysis may
need to be done while concurrently considering the impacts of molecular diffusion
and advection in groundwater pumping wells (Al-Niami & Rushton, 1977). On the
other hand, pollutant mass in groundwater may permeate into bedrock or reduce
the permeability of the soil's pore spaces. Since PFCAs and PFSAs do not break
down, back-diffusion of these low-penetrability composites is likely a more important procedure than that of traditional impurities like chlorinated solvents, and it
could result in PFAS remaining in groundwater for a long time after the source has
been removed and the area has been cleaned up. According to Adamson et al. (2020),
around 82% of the total PFAS quantified at an AFFF site were found in soils with
lower permeability. Ninety-one percent of the polyfluorinated precursor (PCS) mass
was included in this, the majority of which was found close to the suggested source
location. The mass distribution at this location verified the occurrence of diffusion
into soils with reduced permeability and showed how this process might aid in the
long-term retention of PFAS. This study did not thoroughly investigate the relative
influence of PFAS accretion at the air-water interface since the water table was quite
shallow and the unsaturated/saturated transition zone was likely impacted during
excavation. PFAS may be absorbed by concrete and other site materials. Diffusion,
for instance, was identified by Baduel et al. (2015) as a contributory factor in the 12
cm penetration of PFAS into a concrete pad at a fire training site.

Researchers are looking at how diffusion can affect the persistence of PFAS in
natural soils. To understand how this mechanism affects PFAS persistence, suitable
diffusion coefficients for the amount of PFAS that could exist after a release must
be determined. Diffusion coefficients for nine different PFAAs were experimentally
established by Schaefer et al. (2019), who also demonstrated that a decrease in aqueous diffusivity values occurred as the molar volume of PFAS rose. The intricate
chemical interactions of fluorinated compounds caused this relationship to be non-
linear. While the values showed fair agreement with specific comparable techniques
of obtaining diffusion coefficients, they did not agree with all.

Moreover, the differing rates of PFAS diffusion with varying charges—anionic
vs. zwitterionic/cationic—raise other possible problems since porous medium particles may be charged. As previously noted, in the Adamson et al. (2020) investigation,
93% of the mass that had diffused into the lower penetrability regions was composed of zwitterionic and/or cationic polyfluorinated mass; a reduced proportion of
anionic polyfluorinated mass was discovered in these regions. The PFAS's sustained

retention was probably influenced by more OC and beneficial electrostatic inter-actions. By moving PFAS mass to less-transmissive zones, matrix diffusion could improve long-term retention and lower PFAS rates of mass discharge, according to the study's findings.

3.4.2 PFAS Transport via Air

It is known that a wide range of sources emit PFAS into the atmosphere, many of which have been detected in the air. Air plays a vital role in the transportation of persistent organic pollutants (that is, PFAS), allowing them to spread in all wind directions, increasing their global spread, and causing PFAS to be locally deposited in soils and surface waters near the sources of emission (Shin et al., 2012), which may be problematic for site investigations.

The function of air movement is determined by intricate mechanisms, such as vapour-particle PTN, which are specific to per- and polyfluoroalkyl compounds. Aerosols, which are a suspension of solid particles and liquid droplets in the air, can provide a range of environmental media and surfaces that exhibit PFAS PTN behaviour. McMurdo et al. (2008) showed the CCN of PFAS aerosols released from a water surface (where PFAS are frequently found). Because there are several forms of industrial releases (such as stack emissions), airborne transportation of PFAS is a possibly important movement route. Although the mechanisms causing PFAS leaks from industrial sources have not been well investigated, they could include agitating wet surfaces and drying droplets.

Facilities that can release PFAS, as well as other dangerous air contaminants like fluorinated products of incomplete combustion (PICs), into the atmosphere include pyrolyzers, cement, and lightweight cumulative kilns, sewage sludge incinerators, spent carbon reactivation facilities, municipal waste combustors, and soil disorders (Stoiber et al., 2020; Riedel et al., 2021; Krug et al., 2022). There haven't been much thorough waste characterization or large-scale demonstration studies to show how well these treatment facilities remove and eliminate PFAS. A recent study carried out in a full-scale wasted carbon reactivation plant under normal operating condi-tions revealed that all PFAS compounds were removed from the spent carbon and that the reactivation facility's furnace and air contamination management systems destroyed > 99.99% of the PFAS compounds (DiStefano et al., 2022). It could be necessary to evaluate the significance of additional sources, such as wind-blown foam from fire training and response locations or combustion pollutants.

PFAS can exist in gaseous form once in the air or combine with other aerosols or PM suspended in the atmosphere. The industrial processes contributing to emis-sions will determine the gas phase composition. Neutral volatile PCS compounds, or FTOHs, can make up a minimum of 80% of the entire PFAS mass found in ambient air in an urban area and are frequently the predominant PFAS present in the gas stage (Ahrens et al., 2012).

FTOHs also predominate neutral PFAS across open waters and in isolated areas, with nearly all of them existing in the gas phase (Dreyer et al., 2009; Wang et al., 2015; Bossi et al., 2016; Lai et al., 2016). However, ionic PFAS, such as PFOA and

PFOS, are more frequently found in airborne PM and are identified by their low-vapour pressure and high-water solubility. In both urban and semi-rural settings, PFOA is linked to more minor, ultrafine particles, whereas PFOS is linked to more significant, coarser fractions (Ge et al., 2017; Dreyer et al., 2015; Dreyer et al., 2009). The two main methods of removing PFAS from the atmosphere are wet and dry deposition, which might result from gaseous PFAS PTN to water droplets or scavenging of particle-bound PFAS (Hurley et al., 2004; Dreyer et al., 2010). Rain and snow are typical precipitation sources of per- and polyfluoroalkyl substances (PFAS), with wet and dry deposition occurring within a few days (Hurley et al., 2004; Kim & Kannan, 2007; Liu et al., 2009; Chen et al., 2016b; Kwok et al., 2010; Taniyasu et al., 2013; Lin et al., 2014; Zhao et al., 2013). Even 20 years after production was phased down, it has been demonstrated that several PFAS, notably PFOS, are still found in fine PM (PM2.5) in the US, suggesting that there are still sources of these chemicals (Zhou et al., 2021).

Wet and dry deposition are the two forms of atmospheric deposition that are significant for PFAS (Barton et al., 2007; Taniyasu et al., 2013). PFAS can naturally settle, disseminate, or arise on surfaces through various means during dry deposition. They are mostly linked to liquid or particle phases in air (aerosols). Wet deposition is the process wherein precipitation aids in the washout of these aerosols containing PFAS. The two main methods for removing PFAS from the atmosphere are wet and dry deposition, which might result from gaseous PFAS being separated from water droplets or particle-bound PFAS being scavenged (Hurley et al., 2004). Deposition is regarded as an atmospheric sink because it eliminates mass and lowers the possibility of atmospheric movement over longer distances. Thus, the same mechanism may lead to PFAS contamination of aquatic and terrestrial environments. When wind or other physical causes disrupt PM, PFAS that have adsorbed onto soils or other surfaces—mostly indoor surfaces—can be resurrected.

According to Davis et al. (2007), short-range air transport and deposition in terrestrial and aquatic systems close to major emission sources can lead to PFAS contamination of groundwater, soil, and other media of interest. There have been reports of PFAS in groundwater in areas where hydrologic transmission is not a likely cause. Several kilometres have been demonstrated to separate the sources of the pollution, and area hydrology has little bearing on the dispersion patterns of the contamination (Frisbee et al., 2009; Post et al., 2012; Post, 2013; NYSDOH, 2016). Industrial air emissions have the potential to contaminate drinking water supplies and raise the danger to human health in communities located a considerable distance from the source of the emissions (Schroeder et al., 2021). Ionic PFAS emissions from industries are probably related to PM (Barton et al., 2006), which is wet-scavenged by precipitation and falls to the ground in dry weather (Sehmel, 1984; Slinn, 1984). The PFAS air emissions rate, the direction of the prevailing wind, the properties of the soil, the distance from upwind sources, and the depth of the well are essential factors to consider when predicting the effects on groundwater wells (Roostaei et al., 2021). While modelling PFAS in private wells (n~2300) using a statewide data set, researchers discovered that the most important predictor of consequences was proximity to point sources, such as the textile, rubber, and plastics sectors. The quantity of sand and clay in the soil, hydraulic conductivity, precipitation, and groundwater recharge were secondary predictors (Hu et al., 2021).

PFAS deposition has been estimated using predictive algorithms (Shin et al., 2012; D'Ambro et al., 2021). The deposition of gases and aerosols in wet and dry situations is predicted using modules of the AERMOD system, a regulatory model developed by the USEPA and the American Meteorological Society (USEPA, 2016a). The atmospheric fate and transit of PFAS emissions, together with GenX, from a fluoropolymer manufacturing facility in North Carolina have also been investigated using the Community Multiscale Air Quality (CMAQ) model. D'Ambro et al. (2021), found that most emissions are transported over 150 kilometres from the source, with 2.5% of all GenX and 5% of all PFAS by mass being deposited within 150 kilometres or less. Validating the model is crucial, and model projections should be regarded cautiously in cases where ambiguity persists. Nonetheless, the model could aid in understanding how PFAS are distributed in soil and groundwater close to PFAS emission sources (Shin et al., 2012). The height of the release point and surrounding structures, source emission rates and particle size distributions, stack effluent properties, land use characteristics, local topography, and meteorological data are all significant input parameters for emissions from a smokestack or vent. Air models may provide valuable insights into the usage of these models and the interpretation of results, and many states are actively using and analyzing them to evaluate the consequences of PFAS emissions from industrial sources.

Ionic and neutral PFAS have been found in seawater, ice cores, surface snow, biota, and other environmental media in far-flung places like the Arctic and Antarctic. This is conclusive evidence that long-distance transport mechanisms are responsible for the extensive distribution of these compounds across the planet (Ahrens et al., 2010; Cai et al., 2012a, 2012b, 2012c; Benskin et al., 2012; Bossi et al., 2016; Codling et al., 2014; Dreyer et al., 2009; Kirchgeorg et al., 2016; Kirchgeorg et al., 2013; Wang et al., 2014; 2015; Rankin et al., 2016; Yeung et al., 2017; Joerss et al., 2020; Langberg et al., 2022; Szabo et al., 2022). According to geochemical modelling, air inputs may be responsible for anywhere between 34% and 59% of the reported CCNs of PFOA in the polar mixed layer (PML) in the Arctic. Oceanic inputs also play a role in this process (Yeung et al., 2017). The decreases in long-range atmospheric transport caused by the commercial phaseout of specific PFAS have mostly been due to diminishing trends in the CCN of PFSAs in higher trophic arctic species (Routti et al., 2017). PFAS are believed to be distributed to underdeveloped areas far from direct industrial input through a combination of long-range atmospheric conveyance and the subsequent breakdown of volatile PCSs, transfer through ocean currents, and release into the atmosphere as marine aerosols (sea spray) (Ellis et al., 2004; Wania, 2007; De Silva et al., 2009; Casas et al., 2020; Joerss et al., 2020; Lin et al., 2021).

3.4.3 Leaching

When precipitation, floods, or irrigation events occur, they can cause PFAS that are present in unsaturated soils to leach downhill and facilitate the breakdown and movement of the contaminating material (Ahrens & Bundschuh, 2014; Sepulvado et al.,

2011; Sharifan et al., 2021). Since PFAS emissions often include surface applications (such as AFFF and biosolids) or air deposition, this procedure might lead to PFAS transmission from the soil surfaces to groundwater and surface water (Anderson et al., 2019; Borthakur et al., 2022; Galloway et al., 2020; Gellrich et al., 2012).

PFAS migration from shallow soils to groundwater is impacted by a number of interrelated factors that can either increase or decrease the pace at which PFAS leaches. High rates of water penetration, whether from man-made sources like irrigation or natural sources like precipitation, will increase the leaching potential in certain places. Leaching potential is also impacted by the thickness of the unsaturated zone or the distance below the water table. These characteristics match those of other shallow soil pollutants that are not PFAS. This potentially restrict the amount of PFAS that seeps into groundwater from shallow soil. These include PFAS AST at the air-water interface, PTN to NAPL, and PTN to solid phases (such as soil particles). Therefore, soil-to-groundwater leaching may be limited by soil features that increase the potential importance of PFAS-sorptive processes. This entails greater ionic strength, surface area, surface charge, air content, and OC levels inside the vadose zone (Guelfo & Higgins, 2013). The structural properties of the particular PFAS, such as the tendency for longer chains to exhibit more significant PTN coefficients and be less soluble than shorter chains, will also affect their transport efficiency. Furthermore, it has been demonstrated that some of these PTN mechanisms within the vadose zone are non-linear, meaning that when CCNs fluctuate due to dilution and transformation, their proportional contribution to leaching may also alter over time (Zeng & Guo, 2021). Finally, while assessing the possibility of PFAS escaping from soil into groundwater, site features that affect the degree of flushing (such as the rates of precipitation and depth to groundwater) ought to be taken into account.

The pace at which certain PFAS move from the vadose region to groundwater (breakthrough) may be influenced by several circumstances. Similar to hydrophobic PTN, the hydrogeologic and geochemical characteristics of the formation (such as water content and salinity) have a significant impact on the comparative significance of air-water interfacial PTN (Anderson et al., 2019). Generally, air-water interfacial PTN can help retain PFAS in bulk soil, distinct from other organic pollutants. It has been estimated that it will take PFOS (and other PFAS showing high interfacial AST properties) anything from a year to many decades or more to reach the underlying groundwater. Climate and PFOS CCNs will probably affect how much this retardation factor reduces (Brusseau, 2020; Guo et al., 2020). This delay may impact the choice and execution of treatments or even the time that long-term management initiatives last.

PFAS's propensity to be linked to the solid phase and air-filled pore space may cause retention inside the vadose zone, similar to the impact previously mentioned. A small amount of PFAS could be in the aqueous stage and vulnerable to deeper penetration, according to Guo et al. (2020). Due to the possibility of hysteretic desorption from soils with a comparatively high organic matter content due to mass transfer restrictions, solid phase sorption (hydrophobic PTN and electrostatic interactions) as well as air-water interfacial PTN may interact crucially (Schaefer et al., 2021).

PFAS can spread by leaching, according to some experimental and field-based research (Filipovic et al., 2015; Hellsing et al., 2016; Bräunig et al., 2017), others have observed that longer-chain PFAS are retained over time in shallow soils after prolonged percolation (Anderson, 2021; Sepulvado et al., 2011; Stahl et al., 2013). In silty soil with some clay and natural rainfall, PFOA, PFCAs, and PFSAs moved down the soil column more quickly than PFOS, according to a long-term lysimeter study (Thorsten Stahl et al., 2013). Nevertheless, 96.88% and 99.98% of the PFOA and PFOS, respectively, were still present in the soil after five years. This PFOA and PFOS retention might lengthen the (soil-bound) source's lifespan (Baduel et al., 2015). A prolonged groundwater plume might result from the gradual leaching of PFAS from shallow soils, considering the low (part per trillion) amounts in groundwater.

3.5 PFAS TRANSFER ACROSS SEDIMENT, POREWATER, SURFACE WATER, AND GROUNDWATER

Determining the nature of these interactions and their implications for the fate and transport of PFAS is probably site-specific because of the many interactions that PFAS have with sediments, groundwater, pore water, and surface water.

Research conducted in the Little Neshaminy watershed in southeast Pennsylvania, for instance, shows that there are situations in which surface water PFAS CCNs are affected by the groundwater discharge and other conditions in which surface water PFAS CCNs are diluted by the same groundwater discharge (Leidos, 2019; Tech, 2022). Additionally, field studies have shown that PFAS migrate from groundwater to surface water via infiltrating stormwater utilities with groundwater (Leidos, 2019; Wood, 2020). Additional studies have demonstrated that PFAS may move from groundwater to surface features like marshes (SES, 2021). Studies on bigger rivers have shown that seasonal and local fluctuations in stream flow could influence the movement of PFAS from groundwater to surface water by changing the gradients and flow direction of groundwater (Services, 2022).

3.5.1 GROUNDWATER/SURFACE WATER INTERACTIONS

There are sometimes differences in the geochemistry or chemistry of surface water and groundwater, including dissolved oxygen, ionic CCNs, pH, and the sources and quantities of contaminants such as per- and polyfluoroalkyl substances (PFAS). The border layer separating surface water from groundwater is represented by sediment pore water, where PFAS may have significant fate consequences due to geochemical and redox transition zones.

Very few investigations on PFAS at the surface water/groundwater interface have been conducted. Tokranov et al. (2021) examined the dissolved oxygen content, nutrients, and PFAS amounts in the downwelling lake surface water. They discovered that PCS CCNs dropped across the pore water border from surface water to groundwater, and they ascribed this drop to sorption and biotransformation.

It is anticipated that PFAA PCSs will be affected by redox gradients in the sediment pore water zone. The transformation of PFAA PCSs may be altered by microbial population shifts caused by the reaeration of anoxic groundwater during exfiltration to surface water. An equivalent procedure was followed when higher levels of PFAAs were found during field biosparging (McGuire et al., 2014) and column studies (Nickerson et al., 2021). Redox transition zones between surface and groundwater may serve as biotransformation hotspots. Other abiotic changes may occur at the margins between anoxic groundwater and aerobic surface water.

The effect of redox gradients on PFAS sorption is not well understood. However, more mobile products are produced when PFAA PCSs undergo aerobic transformation (Weber et al., 2017), which may have a lower chance of being adsorbed to solid phases. A range of environmental factors, such as pH, aqueous calcium and humic acid CCN, mineral and grain coating composition, and sediment OC content, have been demonstrated to affect PFAS PTN in laboratory experiments, together with the effects on the perfluorocarbon chain length "head group" and PCS transformation (Tokranov et al., 2021). Steffens et al. (2021) discovered that an apparent "salting out" effect caused the observed PFOS CCNs in ionic solutions to drop considerably. Higher salinity was shown to cause a reduction in bulk solution PFOS CCNs due to increased PFOS absorption at the water-container interface and enhanced aggregation at the air-water interface. They speculate that in high-salinity settings or places where salinity fluctuates over time or distance, this effect could affect the transportation of per- and polyfluoroalkyl substances (PFAS). Elevated salinity can cause PFAS to aggregate on particulates, sediment, or other solids and increase their CCNs in surface microlayers.

Determining how much PFAS from groundwater has been introduced to surface waters is challenging, mainly when there are additional surface point discharge sources or when PFAS transit may be significantly impacted by transformation and sorption processes. However, perfluoroether carboxylic acids (PFECAs), which are produced by hyporheic exchange, have been predicted to reach up to 32 kg/yr based on field measurements of surface water and groundwater at a study location that did not have direct PFAS outfalls into the water (Pétré et al., 2021). The authors anticipated minimal sorption and insufficient time for transformation (days to months) because of PFECAs' fast mobility and low retardation factor. They also assumed no sorption or transformation. They suggested that desorption would understate the amount of PFAS absorbed by surface water, whereas AST would overstate it.

3.5.2 SURFACE WATER/SEDIMENT INTERACTIONS

Owing to their high-water solubility, PFAS, especially short-chain PFAS, are mobile in aquatic environments (Campo et al., 2016). This feature makes it more difficult to find a connection between CCNs in water, particularly flowing water, and other media, including sediments, fish, or invertebrates (Campo et al., 2016). Some reported studies have examined the PTN of PFAS between surface water and sediments in lakes and other stationary water bodies, including wetlands and estuaries (White et al., 2015; Mussabek et al., 2019; Bai & Son, 2021; da Silva et al., 2022).

The transfer (PTN) of PFAS from the surface water column or the deposition of PFAS that have been sorbed to suspended particles are two possible explanations for the occurrence of PFAS in sediments. Since partition coefficients recorded in the lab are frequently lower than those reported in the field, it is uncertain what role each channel plays in the relative proportion of PFAS transit to sediments (Zhang et al., 2015a; Li et al., 2018; Rovero et al., 2021). PFAS CCNs in suspended solids from PFAS-affected soil erosion might be noticeably higher than those in suspended solids from equilibrium deposition (Xiao et al., 2019; Borthakur et al., 2021). After being eroded from PFAS-affected soils, these suspended solids mix turbulently with particles in rivers and streams, which may raise the CCN of PFAS in the sediment when they settle.

Estimating the rates and fluxes of PFAS deposition into surface water bodies is possible through the use of dated sediment core analysis. According to a study conducted in the Great Lakes, it could work well for long-chain PFAS that show more sorption to sediment, with CCN at a certain depth interval suggesting deposition there. Short-chain PFAAs were not useful as indicators of the rates of PFAS deposition because they seemed to be more mobile in the sediment column and showed less sorption. The authors (Codling et al., 2018) emphasized that sediment cores may become less helpful in assessing trends in deposition rates when additional short-chain PFAS are employed to replace PFAS applications. In a Swedish investigation, age-dated sediment cores allowed for the computation of PFAS fluxes and deposition rates in ponds impacted by AFFF emissions (Mussabek et al., 2019). Peak deposition rates between 2003 and 2009 were found in the research, and these rates coincided with the activity observed in the vicinity of the water bodies. The study found that seasonal fluctuations in water chemistry, sedimentation, and PTN must be considered when interpreting fluxes (Mussabek et al., 2019). In some environmental situations, resuspension or redistribution of contaminants, and episodic or augmented sediment transport can be caused by heavy precipitation or storm occurrences. Following the storm's passage, resampling of marine sediment locations in Florida revealed that sediment CCNs of total PFAS reduced by 47% on average across nine sites (Ahmadireskety et al., 2021). Sediments can also influence CCNs of surface water. When tributary sediments are transferred to a receiving body of water, it has been demonstrated that they may serve as secondary sources of PFAAs (Balgooyen & Remucal, 2022).

In order to assess surface water CCNs and correlate them with sediment PFAS CCNs, passive pore water samplers may be helpful. This is because sediment PFAS CCNs may result from typical deposition situations of discharges that may be episodic. Rather than producing a CCN snapshot at a certain moment in time, passive sampling produces a time-weighted mean CCN and a buildup of pollutants. Equilibrium devices could not be suitable for intermittent discharges, even though accumulation devices might be able to record the average CCNs in sediments. As a result, the sampler type used needs to be carefully considered.

The mineral content (Fe, Pb, Rb, and As) and PFBS, PFHxS, and PFOS in sediments extracted from Lake Sänksjön in Ronneby, Sweden, were shown to be related, but not to the foc, according to a recent study employing principal component analysis (Mussabek et al., 2020). The relationships between the mineral CCN and other PFAS detected at the site, such as PFOA, PFHxA, and 6:2 FTS, were weaker. Since

PFAS have a unique structure that includes polar and non-polar moieties, their sorption behaviour to sediments relies on hydrophobicity and electrostatic interactions (Ahmadireskety et al., 2021; Lampert, 2018). It is anticipated that the factors influencing soil PTN would also influence PFAS sorption into sediments. Sorption and desorption are expected to directly affect PFAS transport processes in aquatic settings and the hyporheic zone (e.g., particle movement owing to sorption, retention due to sorption, or transport in the dissolved phase).

3.5.3 Transformations

Several investigations have documented the biotic and abiotic changes some polyfluorinated perfluoroalkyl substances undergo. Regarded as PCSs, polyfluorinated PFAS that have been demonstrated to convert usually result in PFAAs. PFAAs haven't been demonstrated to break down or change in any other way in natural environments, though. The main differences that affect transformation potential between polyfluorinated PCSs and perfluorinated products are the presence, location, and number of carbon–hydrogen (C–H) bonds, as well as perhaps carbon–oxygen (C–O) links throughout the alkyl carbon chain. More specifically, a range of biotic and abiotic processes can affect PFAS containing C–H bonds, resulting in shorter-chain PFAAs in the end. Although most of the current research on the biotic and abiotic transformation of polyfluorinated PFAS comes from controlled laboratory experiments (as will be discussed in the following subsection), a growing number of published field studies also highlight the significance of PCSs at a diversity of sites with diverse source setups (Dassuncao et al., 2017; Weber et al., 2017).

3.5.4 PFAA Precursors

PFAAs belong to a broad and varied class of polyfluorinated chemicals, even though they are restricted to a very little quantity of homologous molecules that differ mostly in the length of the carbon chain and the terminal FG. There are believed to be thousands of PFAS on the worldwide market today, most of which are polyfluorinated (Wang et al., 2017a). Though, there is a great deal of uncertainty regarding: (1) the amount of PCS transformation that occurs globally, (2) which environmental sections represent the most transformation, (3) pertinent environmental circumstances that influence transformation procedures, and (4) the rate of transformation and pathways because transformation studies that have been published thus far only cover a small subsample of these PFAS. However, it is reasonable to anticipate an increase in the proportion of total PFAS made up of PFAAs because of transformation over distance, over time, and treatment, both worldwide and (specifically) at contaminated sites.

3.5.5 Atmospheric Transformations

Globally, PFCA direct emissions have been declining, while air emissions of PFCA PCSs have been increasing (Wang et al., 2014; Thackray & Selin, 2017). Similarly,

PFSA PCS emission rates are rising universally (Löfstedt Gilljam et al., 2016). On a regional and global scale, atmospheric movement plays a significant role in the dispersion of per- and polyfluoroalkyl substances (PFAS). As a result, PFAS have been found in isolated areas, including arctic regions. PFAS are also transported to polar regions by ocean currents. However, it's unclear how much each pathway contributes to the other (Yeung et al., 2017). The transportation of PCSs through the atmosphere and their subsequent modification is a significant source of perfluoroalkyl chemicals (PFAAs) in the environment, notwithstanding the comparative contributions of oceanic and atmospheric conveyance.

FTOHs are the main PFCA PCSs in the atmosphere that are often measured (Martin et al., 2002; Thackray & Selin, 2017; Young & Mabury, 2010). Samples of the marine atmosphere were collected during an expedition research trip that travelled through the Atlantic, Arctic, and Southern seas (Wang et al., 2015). The samples were then examined for PCSs, including FTOHs, FTAs, FOSAs, and FOSEs. FTOHs were the most common species, the researchers discovered.

A major source of persistent flame retardants (PFCAs) in the environment, especially in the Arctic, may involve PCS atmospheric changes such as FTOHs (Schenker et al., 2008). PFCA deposition can be greatly aided by indirect photolysis of some PCSs that take place in the atmosphere, even if direct photolysis of PFAS has not been shown (Yarwood et al., 2007; Armitage et al., 2009a). The environmental conversion of 8:2 FTOH to PFOA, for example, occurs through interactions with hydroxyl and chlorine radicals; similar processes occur for 6:2 and 4:2 FTOHs (Ellis et al., 2004); and the potentially atmospheric perfluoroalkyl sulfonamides, which decompose into PFCAs and PFSAs (Martin et al., 2006) and PFSAs (terrestrial environment) (Avendaño & Liu, 2015). Other semi-volatile PCSs, including FTOH, might be atmospherically transformed into PFCAs (Young & Mabury, 2010).

PCSs undergo a multistep atmospheric process that converts them to nitrous oxides (NO_x) and peroxy radicals (RO_2); the yield of the final PFCA product depends on several factors. According to Young and Mabury (2010), isolated locales often have greater long-chain PFCA yields since high NOx levels reduce these yields. For the synthesis of PFOA and PFNA from 8:2 FTOH, Thackray and Selin (2017) determined theoretical maximum yields. Local photochemical circumstances had a substantial impact on these yields, which ranged from less than 1% to 40% (PFOA) or 80% (PFNA).

3.5.6   IN SITU TRANSFORMATIONS

3.5.6.1   Abiotic Pathways

It has been shown that abiotic activities such as hydrolysis, photolysis, and oxidation can alter the PCSs in soil and water when subjected to ambient environmental circumstances. PFSAs can be produced by hydrolyzing specific PCSs and then undergoing biotransformation. One such instance is the discovery of PFOS from perfluorooctane sulfonyl fluoride (POSF) (Martin et al., 2010). PFCAs are produced by other hydrolysis methods. In particular, Washington & Jenkins (2015) showed that monomeric PCSs of PFOA as well as other PFCAs with half-lives of 50 to 90 years at neutral pH are produced by hydrolyzing polymeric PCSs obtained from fluorotelomer. Additionally, PCSs in natural waters can be oxidized by hydroxyl radicals. Fluorotelomer-formed PCSs oxidize more quickly than electrochemical fluorination (ECF)-based PCSs (Gauthier & Mabury, 2005; Plumlee et al., 2009). Perfluorobutane sulfonate (PFBS) and other chain length-reducing PFCAs and PFSAs can also be produced via oxidation processes involving hydroxyl radicals and sulfonamido derivatives (D'eon et al., 2006). Lastly, abiotic PCS transformations may sometimes result in different polyfluorinated sulfonamido intermediate chemicals from ECF-based PCSs without producing any PFAA initially, however, PFAAs may eventually arise.

3.5.6.2 Aerobic Biological Pathways

Research on the composition of PFAS at various stages of wastewater treatment offers evidence of aerobic biotransformation field investigations at AFFF-affected locations, for instance, (Anderson et al., 2016; Houtz et al., 2013; McGuire et al., 2014; Weber et al., 2017), and, most reliably, from studies conducted in microcosms.

In particular, the 6:2 and 8:2 FTOHs in soil, sludge, or aqueous matrices have been the most often examined in microcosm studies. While the reported degradation rates and intermediates differ across different research, the final transformation products, C8 PFCAs, have been consistently detected (Dinglasan et al., 2004; Liu et al., 2007; Sáez et al., 2008; Wang et al., 2009; Wang et al., 2005a; Wang et al., 2005b). However, substantially lower PFCA yields using alternative routes were reported in a pure culture experiment with the white-rot fungus *P. chrysosporium* (Tseng et al., 2014). Other polyfluorinated perfluoroalkylamido sulfonate (6:2 fluorotelomer mercapto alkylamido sulfonate) produced from telomere that have been studied (Weiner et al., 2013), are the 6:2 fluorotelomer sulfonate (Wang et al., 2011), the 4:2, 6:2, and 8:2 fluorotelomer thioether amido sulfonates (Harding-Marjanovic et al., 2015), the 6:2 fluorotelomer sulfonamide alkylamine, the perfluorooctane amido quaternary ammonium salt (Mejia-Avendaño et al., 2016), and the 6:2 fluorotelomer sulfonamide alkylbetaine (D'Agostino & Mabury, 2017). They all show how PFCAs develop at different rates and suggest mechanisms. Numerous investigations have also demonstrated that different ECF-derived polyfluorinated PFAS may be biotransformed aerobically. PFSA PCSs that have been studied include N-ethyl perfluorooctane sulfonamide (Avendaño & Liu, 2015), N-ethyl perfluorooctane sulfonamido ethanol (Rhoads et al., 2008; Zhang et al., 2017), and perfluorooctane sulfonamide quaternary ammonium salt (Mejia-Avendaño et al., 2016). All show PFSA production with varying rates and deduced mechanisms.

3.5.6.3 Anaerobic Biological Pathways

To date, little published research provides clear evidence of the anaerobic biotransformation of per- and polyfluorinated persistent flame retardants (PFAS). Differences have been found in the end products of PFAA PCSs in aerobic and anaerobic circumstances (Choi et al., 2022). While more transformation products are created under more reducing anaerobic settings, including those that minimize iron and sulphate

(Yan et al., 2022; Yi et al., 2018), PFAAs are formed in aerobic and, to some extent, denitrifying environments (Yi et al., 2022). Methanogenic circumstances may result in stable polyfluorinated acids with significantly slower KNT than aerobic biotransformation, as demonstrated by two examples involving FTOHs which were examined (Allred et al., 2015; Zhang et al., 2013). According to a recent study, a specific microorganism strain (Acidimicrobium sp. Strain A6) defluorinates PFOA and PFOS under Feammox conditions, which decrease iron and oxidizes ammonium (Huang & Jaffé, 2019). The environmental effects of anaerobic biotransformations of polyfluorinated PFAS as PFAA sources may or may not be significant.

3.5.7 Polymer Transformation

Fluoropolymers, side-chain fluorinated polymers, and polymeric perfluoropoly-ethers are all groups of the PFAS family of polymeric chemicals (Buck et al., 2011). According to Henry et al. (2018), fluoropolymers are high-molecular-weight solid plastics (> 100,000 Daltons, or Da) in which the fluorine atoms are directly bound to the carbon atoms in the carbon polymer's backbone. F is directly bound to carbon atoms in the ether polymer backbone of polymeric perfluoropolyethers. The primary applications for these complex polymeric PFAS are surface protectors and surfactants. Telomere-derived PCSs are used to make side-chain fluorinated polymers, which have fluorinated side chains and a non-fluorinated polymeric backbone.

In many commercial and industrial applications, side-chain fluorinated polymers are utilized as surfactants and chemicals to protect surfaces (Buck et al., 2011). It is crucial to comprehend these polyfluorinated polymers' capacity for biotransformation. Nevertheless, little research has been done on the possible biotransformation of side-chain polymers, such as the urethane polymer based on fluorotelomer (Russell et al., 2010). Considering the intricacy of side-chain fluorinated polymers, these investigations differ significantly. The main issue is that polymer CCNs cannot be monitored. Due to the lack of analytical techniques for the direct quantification of polymers, all investigations other than Rankin et al. (2014) observed potential FTOH breakdown products as opposed to the polymer's disappearance (Dasu et al., 2012; Dasu & Lee, 2016; Liu et al., 2007; Wang et al., 2009, 2005b. Rankin et al. (2014) used matrix-assisted laser desorption/ionization time-of-flight (MALDI-TOF) mass spectrometry to track the polymer's disappearance and recognize breakdown products qualitatively.

Furthermore, contaminants or non-polymerized leftovers (such as monomers, oligomers, PFCAs, FTOHs, etc.) make it more difficult to interpret data and may skew findings about polymer biodegradation. Last but not least, the two-year maximum length of the biodegradability tests is far less than the side-chain fluorinated polymers' anticipated half-lives, which range from decades to thousands of years. Modelling assumptions are, therefore, also important sources of uncertainty.

The potential of the biodegradation of two distinct side-chain fluoropolymers in soils—urethane polymer and fluorotelomer-based acrylate polymer—was studied for two years (Russell et al., 2010). Based on the experimental results for PFOA, the projected half-lives of urethane and acrylate polymers were 1,200–1,700 years and

28–241 years (with a geometric mean of 102 years), respectively. However, the polymer used in this investigation contained notable residuals. In addition, Washington et al. (2009) looked at the biodegradation capability of polyacrylate based on low-residue fluorotelomers. Based on PFOA experiments' findings, the acrylate polymer's half-life was estimated to be between 870 and 1,400 years.

Additionally, the researchers simulated and predicted the half-life for finely grained polymers, which are frequently found in commercial goods, assuming that degradation is surface-mediated. By normalizing to the anticipated surface area of the polymer, they were able to determine a half-life of 10–17 years. This suggests that one of the foremost sources of PFCAs in the environment may be side-chain fluoropolymer compounds with small granules. Using comprehensive extractions, Washington et al. (2015) examined the biodegradability of commercial acrylate polymer in soils over a period of 376 days, with estimated half-lives ranging between 33 and 112 years (Washington et al., 2014). This investigation also revealed that the acrylate polymer degrades ten times more quickly in pH 10 water when subjected to OH-mediated hydrolysis than when treated neutrally. More research is necessary to determine the abiotic degradation mechanism because there was hardly any abiotic transformation of a side-chain fluorinated polymer documented in the available literature. Another study team, Rankin et al. (2014), examined the biodegradation of an acrylate polymer based on fluorotelomer in soil, plants, and biosolids over a period of 5.5 months. Compared to soils, plants and biosolids degraded more quickly. A wide range of predicted half-lives, ranging from 8 to 111 years, were also discovered by this investigation. The modelling assumptions used in various investigations have resulted in varying half-lives (Russell et al., 2008, 2010; Washington et al., 2009, 2010, 2015, 2019).

3.6 PRACTICAL IMPLICATIONS

PCS transformation should be considered during thorough site inspections, as it might complicate CSMs and risk assessments (RAs). For example, even in the absence of obvious point sources, long-distance transport brought on by air emissions of volatile PCSs may cause noticeable levels of PFAAs in ecological media through subsequent transformation and deposition (Vedagiri et al., 2018). Furthermore, because of PCS transformation, events such as oxidation, downstream CCNs of PFOA and PFOS in groundwater, or downstream CCNs inside treatment and management systems are frequently more significant than those near the source region or influent. Additionally, when PCSs change over time and space, PFOA and PFOS CCNs may also show a rising trend. This might impact the scope and data evaluation of the Remedial Investigation/Feasibility Study (RI/FS).

When PFAS that would not otherwise be found are transformed into detectable PFAAs, it is crucial for site studies into PCSs associated with the site, provided that the change takes place after the first site characterization attempts. Furthermore, it has been proposed that the differing transit rates of PCS PFAS and the associated terminal PFAA might deceive CSMs if transformation rates are more sluggish than conveyance rates (Weber et al., 2017).

Surrogate analytical techniques have been developed to account for PCSs that would otherwise be impossible to measure. These include the total oxidizable PCS (TOP) test (Houtz & Sedlak, 2012), adsorbable organic fluorine (AOF), followed by ion chromatography combustion (Wagner et al., 2013), and spectroscopy using particle-induced gamma-ray emission (PIGE) (Schaider et al., 2017). Here you may get more details about these additional analytical techniques for figuring out PCS CCNs .

3.6.1   PFAS Uptake into Aquatic Organisms

There is a tendency for certain PFAS to bioaccumulate. In other words, they are absorbed by organisms and stored in environmental media. Lower CCNs of these chemicals have been found in fish, lower trophic level invertebrates, and the tissue of species at the base of the food chain. Additionally, certain PFAS have the ability to biomagnify up the food chain. Predator fish and air-breathing species at the top of the food chain, such as polar bears and seagulls, have been shown to have PFAS levels (Houde et al., 2011; Gobas et al., 2020a; Burkhard, 2021). Air-breathing creatures (such as aquatic-dependent birds and mammals) seem to have more substantial trophic transfer of PFAS and biomagnification than gill-breathing species (like higher trophic-level fish species). This is because PFAS are eliminated from the respiratory system much more readily through the gills to water than lungs to air (De Silva et al., 2021). Note that not all PFAS biomagnify, and there is conflicting data for some (like PFOS) in this regard (Franklin, 2016).

The two main processes regulating the bioaccumulation of PFAS in aquatic environments are absorption, which occurs from food and water; and depuration, which involves excretion through the gill surface, urine, and faeces (Chen et al., 2016a; Zhong et al., 2019), evolution dilution, as well as biotransformation. The degree of accumulation varies across different types of tissues (liver, muscles, whole or entire body, etc.) within an organism.

Ionic PFAS bioaccumulation in aquatic biota is typically linked to proteins, in contrast to non-ionic polar chemical compounds that build up in fatty tissues. Numerous investigations have shown that PFOS and PFOA are frequently associated with serum albumin along with liver and kidney proteins (Han et al., 2003; Jones et al., 2003). These findings suggest that alterations in the kinds of proteins found in each tissue may account for a significant amount of the variation in PFAS between organs (Ng & Hungerbuehler, 2015; Ng & Hungerbü̈hler, 2014). Phospholipid binding is another way PFAS can accumulate (Dassuncao et al., 2019). Therefore, approaches and models that use lipid normalization to assume PTN to lipids to describe the bioaccumulation of non-ionic polar chemicals are probably not relevant for PFAS.

To understand the nature of bioaccumulation, for instance, the European Union's REACH classification of composites as "bioaccumulative" in the "Persistent, Bioaccumulative, and Toxic" (PBT) and "Very Persistent and Very Bioaccumulative" (vPvB) assessments make use of BCFs, BAFs, and BMFs. They also support the development of RAs and surface water quality standards. PFAS BCFs, BAFs, BMFs, and TMFs have been the subject of too many papers for this study to cover them all.

Numerous papers have studied the bioaccumulation of PFAS (Martin et al., 2013; Gobas et al., 2020; Burkhard, 2021; A. O. De Silva et al., 2021).

Based on a variety of criteria, Burkhard (2021) classified each research into one of three data quality groups and constructed a parallel, mostly overlapping database of BAF values. However, there was no quantitative ranking in this case. The distributions show that there aren't many variations between the three measurement quality scores, the authors found (Burkhard, 2021).

Some sources of information on recommended procedures for carrying out laboratory bioaccumulation testing are ASTM, 2013; OECD, 2012; USEPA, 2016b. For instance, data on the planning and analysis of field and lab BAF research are also accessible from Burkhard, 2003; Gobas et al., 2020b. These resources help direct how the research in the database is interpreted.

3.6.2 BIOCONCENTRATION

Research has shown that their (PFAS) chemical structure affects their propensity for bioaccumulation, namely the length of the perfluoroalkyl chain and the presence of FGs (sulfonate vs. carboxylate). Martin et al. (2003) exposed rainbow fish for over 12 days to individual PFAS. BCFs were low for PFAS with shorter perfluoroalkyl carbon chain lengths (less than 7 for carboxylates and less than 6 for sulfonates). It has been shown that the absorption of PFAS compounds is inversely correlated with the CMC, or the CCN at which half of the molecules are linked as MCs, and proportionate to the length of the carbon chain at 8–12 carbons. Additional research has revealed that it is challenging to gather and bioconcentrate shorter-chain PFSAs and PFCAs (Houde et al., 2011; Martin et al., 2013).

Martin et al. (2003) also demonstrated that PFSAs had higher BCFs and half-lives than comparable PFCAs with the same chain lengths, indicating that the FG needs to be considered and that hydrophobicity—as projected by the CMC—is not the only feature affecting the potential for PFAS bioaccumulation. In terms of tissue categories, Martin et al. (2003) demonstrated that the most common locations for PFAS collection were the blood, kidney, liver, and gall bladder. Following the gonads, smaller quantities are deposited in muscle and adipose tissue. PFAS's strong affinity for binding to phospholipids (Dassuncao et al., 2019), fatty acid-binding proteins, and serum albumin (Ng & Hungerbüˊhler, 2014) is probably the reason for the tissue-dependent dispersion seen in several previous investigations.

3.6.3 BIOACCUMULATION

As demonstrated, BAF CCNs recorded in entire bodies or particular organs can be used to report BAFs. The BAFs provided for fish fillet or muscle help to define corresponding water quality parameters and assess the danger of fish ingestion to human health. Entire-body BAFs could possibly be utilized for environmental RAs and setting suitable water quality requirements for higher trophic species (e.g., predatory fish, piscivorous birds, and mammals). Nevertheless, Conder et al. (2020) recommended employing lab-based BCFs as an alternative to field-based BAFs for environmental RAs, mainly due to the fact that lab-based BCFs are less unpredictable and more consistent than field-based BAFs. Moreover, PFAS linked to recently spiked lab medium could be more bioavailable than in the field, resulting in conservative BCFs (greater than what could happen in the wild). PCSs may impact field-based BAFs (Langberg et al., 2020), resulting in the unreliability of risk evaluations and water quality standards based on these BAF calculations. However, in some circumstances, such as when developing site-specific water quality standards and refining site-specific RAs, it may be crucial to consider the site specificity of field-based BAFs.

It is necessary to look at the variations in the degree of accumulation across tissues when comparing BAFs between species or creating statistical summaries of BAF data. As tissue-specific BAFs might change between species due to variations in toxic-KNT, the easiest and most reliable method for comparing BAFs between organisms is to provide BAF values on an entire-body basis. Such data can be acquired by either entire-body samples or tissue-specific measurements that may be statistically integrated to produce equal entire-body CCNs. PFAS CCNs in aquatic creatures' muscles are sometimes twice as high as in their entire bodies (Goeritz et al., 2013; Shi et al., 2015).

Numerous studies examining PFAS residues in wildlife have revealed that certain PFAS have bioaccumulated in various animals worldwide, mostly fish-eating species. There are currently bioaccumulation statistics available for several contexts, such as metropolitan regions and places remote from particular sources, companies that utilize PFAS in production procedures, firefighting, wastewater treatment plants, and water bodies downstream or close to PFAS producers. Most of the sampling has been done to support RAs, advice for eating fish, and theories regarding long-term destiny and travel, temporal patterns, and source identification. The majority of research has been conducted in the Northern Hemisphere, although several writers have discovered that the Southern Hemisphere contains lower amounts (Ahrens, 2011; Ahrens et al., 2009; Armitage et al., 2009). However, according to a study in South Australia's highly industrialized areas, dolphins' livers had some of the highest perfluorocarbons (PFOS) recorded worldwide (Gaylard, 2017).

The following are some observations of PFAS bioaccumulation of PFCAs, PFSAs, and PCSs: The carbon chain length affects PFAA bioaccumulation in a manner similar to bio-CCN (Brendel et al., 2018; USEPA, 2017a). It is believed that PFSAs with less than 6 perfluorinated carbons (PFBS, for example) and PFCAs with fewer than 7 perfluorinated carbons (PFCAs shorter than PFOA) are less bioaccumulative. Compared to PFCAs with the same carbon chain length, PFSAs are more bioaccumulative. The field-measured bioaccumulation and biomagnification factors exhibit significant variability (Burkhard, 2021; Franklin, 2016; Gobas et al., 2020a). Certain PFAAs have BAFs that fall within the range that is frequently utilized as a bioaccumulation criterion, such as the 1,000–5,000 L/kg range used by the USEPA, Canada, and the European Union.

When PFOA and PFOS BAFs for all species are compared, it is evident that PFOA has substantially lower BAFs than PFOS. Additionally, thorough evaluations (Ahrens & Bundschuh, 2014; Houde et al., 2008) reveal that, although PFOS and

PFOA can have similar amounts (~1–10 µg/kg) in invertebrates, PFOS (8-carbon chain PFSA) is usually the more frequent PFAS seen in fish and other air-breathing animals. Ahrens & Bundschuh (2014) claimed that the shorter chain was the reason for the reduced bioaccumulation of PFOA compared to PFOS. The results show that BAFs for perfluorocarboxylates rise to 11 or 12 perfluorinated carbons before declining for more prominent compounds (Langberg et al., 2020).

Lastly, it is critical to remember that PCSs of PFAS could add to the body burden of PFAS. Asher et al. (2012) and Langberg et al. (2020) supplied field data showing that PCSs play a significant role in internal transformation-induced increases in PFOS CCNs in organisms. Furthermore, tests of the atmosphere have demonstrated the ubiquitous presence of PCSs of perfluorinated sulfonamide alcohols (FTOHs) and other PFAS compounds. The PCS or PCSs may be converted into PFOS (from N-ethyl perfluorooctane sulfonamidoethanol, for example) or PFOA (from 8:2 fluorotelomer alcohol, for example) after being ingested by a creature (Galatius et al., 2013; Gebbink et al., 2015). More studies are necessary on the potential role of PFAS PCSs in fish and animal bioaccumulation.

3.6.4 BIOMAGNIFICATION

Research on PFAS in polar and marine settings, particularly the Great Lakes, has generally shown that trophic-level biomagnification within a food web is possible, mostly for PFOS as well as some long-chain PFCAs (Martin et al., 2004; Tomy et al., 2004, 2009; Butt et al., 2010; Houde et al., 2006, 2011).

BMFs for a pelagic food web in Lake Ontario, where lake trout are the predominant predator, were computed by Martin et al. (2010). After accounting for benthic vs. pelagic animals, they demonstrated the biomagnification of some PFAS chemicals, with TMFs ranging from 0.51 for FOSA to 5.88 for PFOS.

The levels of PFOS and C8–C14 perfluorinated carboxylates in bottlenose dolphins in two maritime settings (Charleston, SC; Sarasota Bay, FL) were examined by Houde et al. (2006). Based on estimated TMFs, Franklin (2016), these scientists, who are also referenced, concluded that PFOS and C8–C11 PFCAs are biomagnified in this marine diet. Interestingly, PFOA showed the highest TMF of all the PFCAs, with values gradually declining as chain length amplified. Conder et al. (2008) discovered similar outcomes for perfluorinated acids, with BMF values varying from 0.1 to 20 with a geometric mean of two. They came to the conclusion that the length of the fluorinated carbon chain can be directly connected to the bioaccumulation of ubiquitous lipophilic compounds (PFCAs) (PFSAs with fewer than 6 carbons and PFCAs with below 7 carbons do not biomagnify), which was validated by Lescord et al. (2015). Only PFOS, a perfluorinated acid, regularly exhibits the ability to biomagnify, according to Conder et al. (2008), and the biomagnification of PFCAs in aquatic food webs is lesser than that of the majority of persistent lipophilic composites. Last but not least, Butt et al. (2008) identified PFAS biomagnification factors in the "ringed seal-polar bear" food webs in the Canadian Arctic. PFOS, PFOSA, and C8–C14 PFCAs all had biomagnification factors more than one. Like Houde et al. (2006), they noticed that when the number of carbon chains rose, BMF decreased.

Lescord et al. (2015) found no evidence of biomagnification in the combined levels of PFOS, PFCA, PFSA, and PFAS, in contrast to prior freshwater research. The quantities of the PFAS compounds under study (PFOS, total PFCA, total PFSA, and total PFAS) in five of the six lakes showed a negative connection with trophic level, as determined by stable nitrogen isotope ratios (δ15N). This suggests that biomagnification is not occurring in these freshwater arctic food webs. Overall, their findings demonstrated that the PFAS contents of a taxon are influenced by its horizontal, but not vertical, location in the food web.

Depending on the kinds of tissues examined and the researcher's assumptions about biomagnification in connection with the prey's diet (often ascertained by examining the contents of the stomach), calculated BMFs might vary considerably. This is due to the fact that the BMF or TMF is calculated by dividing the CCN of a predator by the CCN of its prey. Franklin (2016) examined the findings of 24 peer-reviewed papers that reported BMFs or TMFs for 14 PFAS derived from the field. The BMF values varied from 0.01 to 373 (inclusive of only non-zero values), spanning many orders of magnitude. TMFs (containing just non-zero values) ranged from 0.1 to 20. Franklin (2016) ascribed this variability to several causes, such as variations in the metrics' expression (individual vs. entire-body analysis, for instance), failure to attain the presumptive steady-state circumstances, ambiguities in the feeding ecology, and the metabolism of PCS substances.

3.6.4.1 Bioaccumulation Characterization in the Advancement of Standards and Risk Evaluations

The appropriate method for figuring out the bioaccumulative status of PFAS and doing RAs is still up for debate since there is a dispute over the degree of bioaccumulation and/or biomagnification of various PFAS, most notably PFOS (Conder et al., 2020; Franklin, 2016). Franklin (2016) maintained that "field-measured BMFs and TMFs are of very restricted usefulness for assessing bioaccumulation potential status because, in practice, the study-to-study (and even within-study) variability of the results is so great". Franklin (2016) recommended that laboratory BCFs and BMF measurements conducted under close supervision be the only methods used to assess a drug's bioaccumulative status, instead of depending on field-measured BMFs or TMFs. Studies on fish could follow the OECD 305 procedure (OECD, 2012). Dietary BMF assessments may be conducted for terrestrial and avian species utilizing laboratory rats, cows, pigs, or birds, given naturally contaminated feed.

In a similar vein, Conder et al. (2020) proposed that laboratory-derived BCFs be utilized instead of field-based BAFs in ecological RAs because the former may be more reliable and less variable overall. Furthermore, as PFAS may be simpler to access in recently spiked ambient medium than in aged PFAS in field samples, lab-based BCFs are anticipated to yield conservative estimates of bioaccumulation. Moreover, PFAS PCSs that can transform into stable PFAS in field samples do not pose any issues for laboratory BCFs, according to Conder et al. (2020).

Nevertheless, methodological restrictions exist on laboratory BCFs (research where organisms are exposed to contaminants solely in water), such as with very sorptive chemicals. Additionally, because laboratory BCFs do not consider biomagnification, they typically yield poor estimates of actual BAFs. A method similar to the one proposed by Franklin (2016), merging BCFs measured in a lab with BMFs recorded in a controlled environment, could make sense.

Furthermore, laboratory BCFs do not take into consideration the contributions of PCSs to tissue CCNs of PFOS, even if they are able to fix the problems with PCSs. The highest PFOS BAFs as Langberg et al. (2020) determined, with PFOS in yellow perch reaching a maximum of about 2.50×10^5 L/kg wet weight muscle. This result is significantly higher than values often observed in the range of 1,000–10,000 at other sites. The reason for this high number was that the tissue CCNs of PFOS in organisms at this site are mostly determined by PCSs. The true tissue CCN in an ecological risk assessment for such a location would be greatly understated if it were calculated using observed water column PFOS CCNs and a laboratory-based BCF. Therefore, measuring the drug in water is inappropriate in the first place if PCS levels in the system predominantly dictate tissue CCNs. Instead, it is necessary to quantify the organism's exposure to the PCS(s) itself.

Gobas et al. (2020a) suggested that the best way to assess a compound's bioaccumulative tendency is to quantify the BMF utilizing dietary-based OECD 305 research (OECD, 2012), and the BCF and BAF were estimated using a two-compartment bioaccumulation model. The range of applications for which the model created by Gobas et al. (2020) was restricted to non-ionic organic compounds, necessitating modification for use with PFAS.

According to Houde et al. (2008 and 2011), linear PFAS are thought to be more bioaccumulative than branched PFAS, and there may be variations in the rates of PFAS bioaccumulation between the two isomers (Conder et al., 2020). A field population exposed to a variety of isomers would have a risk assessment that is cautious but maybe unrealistic if more bioaccumulative isomers were used in laboratory BCF experiments. Finding the relative abundances of isomers in the environment and assessing the significance of bioaccumulation changes are essential for accurately assessing the influence of isomer composition.

The biases and limitations connected with laboratory BMFs and BCFs should be balanced against the inconsistency of field-measured BAFs and BMFs to ascertain bioaccumulative status and carry out RAs. PFAS' potential to confuse must be considered while doing these assessments, compounded by the compound. Given the uncertainty surrounding PFAS bioaccumulation, a weight-of-evidence approach is probably warranted.

3.6.5   PFAS Uptake into Plants

Plants are predicted to absorb part of the HDP FG found in PFAS via their root systems, whereupon they will be translocated to stems, shoots, leaves, and fruiting bodies. It would be predicted that plants growing nearer polluted sources or watered with PFAS-containing water will collect more significant quantities of PFAS than plants growing farther away from the source (Gobelius et al., 2017), depending on site-specific factors like the characteristics of the soil. Biosolids applied to land have been shown to contaminate the soil with PFAS, and animals fed silage from land-applied fields may have tissues containing higher than normal levels of PFAS (Lindstrom et al., 2011; Skutlarek et al., 2006). Concerns of adulteration of wildlife that eats plants from agricultural regions are also brought up by this. According to the notion that bio-CCN in the latter may happen through the stomata, it was demonstrated that airborne PFAS emissions from Chinese industrial locations impacted the quantity of PFAS in bark and tree leaves (Shan et al., 2014).

Assessing the fate of PFAS throughout different environmental compartments, especially in food chains, with consequences for ecological and human exposures to PFAS, requires an understanding of how plants absorb PFAS. Consumption of vegetables contaminated with PFAS and dairy and meat products from animals given plant-based feed can expose humans to these chemicals. Plants in damaged areas may absorb PFAS and release them into the environment, potentially exposing animals to PFAS.

3.6.6 Uptake Pathways

A thorough analysis of PFAS absorption and accumulation in plants is given by Wang et al. (2020). The overview provided in this section is based on that study. Although to varying degrees, PFAS absorption across a range of natural and cultivated species has been thoroughly investigated, as Wang et al. (2020) report. While many additional PFAS have been demonstrated to be susceptible to plant absorption, the bulk of this research has concentrated on the uptake of PFOA and/or PFOS by crops. Wang et al. (2020) revealed that at least 16 field studies were found, even though the majority of published investigations were carried out in well regulated laboratory settings. These field studies have often engrossed on the point sources of PFAS, such as landfills, wastewater treatment plants, manufacturing complexes, and fire training grounds.

PFAS and soil conditions affect plant uptake. PFAS with lower soil/air-water interface (AWI) retention affinity and greater aqueous solubilities/diffusivities are often found to have improved absorption potential (bioavailability). A bio-CCN factor, for instance, is the ratio of the CCN of PFAS in soil to that in plant tissue. Wang et al. (2020) collected the bio-CCN factors (BCFs) for diverse PFAS that had been described in research articles in the literature. The median BCF values for PFBA, as well as other short-chain PFAS were, on average, more than ten times higher than those for long-chain PFAS, such as PFOA and PFOS, according to this collection. These trends confirm that the degree of plant absorption is influenced by the physical-chemical properties of PFAS. Similarly, physical-chemical soil characteristics (e.g., pH, salinity, temperature, and organic matter content and composition) influence plant absorption of PFAS. But this hasn't been looked at in detail. Additionally, as will be covered in following subsections, plant species and physiology—including transpiration rate and protein content—are significant factors.

Root absorption from soil and water is the main mechanism by which PFAS accumulate in plants (Stahl et al., 2009; Wen et al., 2013; Lee et al., 2014). Additionally, PFAS have been shown to be aerially absorbed from the surrounding environment (both particle-bound and vapour-phase), for example, into bark and leaves (Stahl et al., 2014; Jin et al., 2018; Tian et al., 2018; Liu et al., 2019). However, aerial absorption contributes very little to the overall accumulation of PFAS in plants (Wang et al., 2020).

As previously mentioned, the interactions with the soil phases and the AWI that regulate the quantity of PFAS accessible in soil porewater for root absorption are mostly caused by the aqueous solubility of particular PFAS. Because of a local CCN gradient, PFAS in soil porewater diffuse and transpire their way towards plant roots (Lechner & Knapp, 2011). According to recent research by Zhang et al. (2019) and Wen et al. (2013), cell membrane transport proteins such as anion channels and aquaporins provide a CCN-dependent root uptake mechanism. PFAS penetrate the root epidermis, cortex, and endodermis through apoplastic and symplastic routes before entering the vascular tissue (Blaine et al., 2013).

When they enter the root xylem, PFAS are transported to many plant components, including the stem, shoots, leaves, fruits, and grains. The transpiration stream appears to influence the degree of PFAS translocation in these tissues, with more PFAS accumulation taking place in areas that have higher sorption or incorporation capacities and receive higher water intake (Lechner & Knapp, 2011; Krippner et al., 2015; Gobelius et al., 2017). For example, Lechner & Knapp (2011) found that the accumulation of PFOA and PFOS was lower in peeled edible portions of potatoes, carrots, and cucumbers grown on soil contaminated with PFAS-infected sewage sludge than in foliage, leaves, and stalks. Compared to spring wheat, oat, and maize straw, Stahl et al. (2009) found significantly higher levels of PFOA and PFOS in grains. According to Gobelius et al. (2017), who looked at the distribution of PFAS in trees at an AFFF release site, the overall PFAS accumulation in birch and spruce trees was distributed as follows: leaves > twigs > trunk/core or roots.

3.6.7 BIOCONCENTRATION/BIOACCUMULATION

BCF and BAF are used interchangeably in the plant uptake studies described previously. To calculate them, the PFAS CCN in the plant (mass/mass) is divided by the PFAS CCN in the soil (mass/mass). Research where PFAS were sprayed on crops by irrigation or soil treated with biosolids produced several BAF values (Blaine et al., 2013; Blaine et al., 2014a, 2014b). Inedible crop components (like plant leaves) and edible (such as fruit, lettuce leaves, and roots) were gathered for the study. Other BCFs and BAFs were discovered in investigations of plants exposed to soil, groundwater, surface water, or air containing per- and polyfluoroalkyl substances (PFAS) at PFAS release sites (Mudumbi et al., 2014; Zhang et al., 2015b; Gobelius et al., 2017). It is commonly known that 1) shorter-chain PFAS are more easily absorbed than longer-chain homologues, and the majority of plant BCFs and BAFs fall between 0.1 and ten. There is no net PFAS buildup from soil to plant when the BCF or BAF is 1.0. A BCF or BAF of this type shows that the CCN of PFAS per unit weight in the soils and target plant is the same. This is not to say that soils and plants are in harmony. Certain vegetables, such as lettuce, have a high water content, which may explain why the BAF of 56.8 was so high (Blaine et al., 2013). In controlled food crop studies, short-chain PFCAs and PFSAs have greater BAFs than long-chain composites.

Pervasive formative acidity (PFAA), greenhouse lettuce, and tomato growth in soils that were modelled to resemble control soil, municipal biosolids-amended soil, and industrially affected biosolids-amended soil (but incorporating contaminated biosolids ten times higher than the agronomic rates permitted for Class B biosolids) were investigated by Blaine et al. (2013). The BCFs for the different PFAAs in the industrially impacted biosolids-amended soil were significantly higher than unity; PFPeA had the greatest BCF in tomato (17.1) and PFBA had the highest in lettuce (56.8). BAFs for PFCAs and PFSAs were usually a little higher ($\sim$0.3–0.8 log units) in the industrially impacted soil than in the municipal soil. The BCFs for PFAAs in greenhouse lettuce decreased by 0.3 log units for every -CF2 group. Additionally, they collected soil, corn stover, and corn grains from several large-scale agricultural sites that received nitrogen amendments from biosolids at the agronomic rate. All PFAAs were below the LOQ, and very little amounts of PFPeA and PFBA were found in the maize stover in these regions. According to research by Blaine et al. (2013), crop type, soil characteristics, analyte, and PFAA CCNs all have a substantial impact on the bio-CCN of per- and polyfluoroalkyl compounds (PFAAs) from soils treated with biosolids.

Ghisi et al. (2019) gave an overview of the potential for PFAS absorption from soil, groundwater, and air in agricultural plants. Plant species, growth media, soil organic matter, PFAS chain length, FG, and other soil characteristics influence plant absorption. Increased PFAS CCNs in soil are often associated with increased PFAS CCNs in plants, while there may not be a clear correlation between the two. While long-chain chemicals tend to concentrate more on roots, short-chain compounds collect at higher CCNs in fruits and leafy crops. PFCAs build more quickly than PFSAs, according to several studies. According to several reviewed studies, soil organic matter can trap PFAS and restrict plant absorption. An additional analysis of agricultural plants' bioaccumulation components is available from Lesmeister et al. (2021).

Gobelius et al. (2017) investigated the absorption of 26 PFAS by plants, namely trees, in an AFFF (fire training) site where soil and groundwater were contaminated. Tissues (roots, spruce, cherry, ash, elder, beech fern, leaves/needles) and various plant species (wild strawberry, birch, spruce, cherry, and ash) were gathered with groundwater samples. Out of all tissues, foliage exhibited the greatest BCFs, ranging from 0 to 14,000, and the most significant amount of PFAS (8 out of 26). Birch sap, in particular, showed BCF values as high as 41 for 6:2 FTSA. One possible explanation for the highest mean BCFs for 6:2 FTSA (472; n = 52), PFOS (28; n = 36), PFHxS (10; n = 42), and PFOA (5; n = 24) could be the FTSA composition used at the establishment. Cherry (0.25 ± 0.043) and birch (1.2 ± 1.5) were closely behind spruce, which had mean BCFs (±s.d.) for PFOA of 18 ± 15. All plant species contain PFAS, according to the scientists' findings, and the chemicals' distribution followed the "shoots to roots" pattern: leaves > twigs/stems > trunk > roots. To bolster their assertion that "this order has proven applicable to all samples and species", Gobelius et al. cited additional publications. As a result, PFAS often build up in the plant's vegetative sections as opposed to its storage tissues.

3.7   CONCLUSION

Addressing the persistence and possible impact of PFAS requires understanding their environmental fate and mobility. The complex processes of PFAS migration and PTN across environmental compartments are highlighted in this chapter. The distribution and mobility of PFAS are greatly influenced by their interactions with solid phases, organic matter, and air/water interfaces. Their environmental dynamics are further complicated by diffusion via lower-permeability materials, hysteresis, and non-linear sorption behaviours. Furthermore, the interrelated routes that support PFAS persistence and bioavailability are revealed by interphase transfers, such as those that occur between sediment, porewater, surface water, and groundwater. Researchers and policymakers may create more focused remediation and risk management plans and more accurately forecast PFAS behaviour in various contexts by expanding our understanding of these processes. To address the pervasive and long-lasting environmental impact of PFAS, future research should emphasize improving PTN models and investigating novel approaches for PFAS abatement.

REFERENCES

Adamson, D. T., Nickerson, A., Kulkarni, P. R., Higgins, C. P., Popovic, J., Field, J., Rodowa, A., Newell, C., DeBlanc, P., & Kornuc, J. J. (2020). Mass-based, field-scale demonstration of PFAS retention within AFFF-associated source areas. *Environmental Science & Technology, 54*(24), 15768–15777.

AECOM. (2021). *Surface water foam study report.* https://www.lambeth.gov.uk/sites/default/files/2022-06/Lambeth-Surface-Water-Management-Plan-2021.pdf

Ahmadireskety, A., Da Silva, B. F., Awkerman, J. A., Aufmuth, J., Yost, R. A., & Bowden, J. A. (2021). Per-and polyfluoroalkyl substances (PFAS) in sediments collected from the Pensacola Bay System watershed. *Environmental Advances, 5,* 100088.

Ahrens, L. (2011). Polyfluoroalkyl compounds in the aquatic environment: A review of their occurrence and fate. *Journal of Environmental Monitoring, 13*(1), 20–31.

Ahrens, L., & Bundschuh, M. (2014). Fate and effects of poly- and perfluoroalkyl substances in the aquatic environment: A review. *Environmental Toxicology and Chemistry, 33*(9), 1921–1929.

Ahrens, L., Barber, J. L., Xie, Z., & Ebinghaus, R. (2009). Longitudinal and latitudinal distribution of perfluoroalkyl compounds in the surface water of the Atlantic Ocean. *Environmental Science & Technology, 43*(9), 3122–3127.

Ahrens, L., Harner, T., Shoeib, M., Lane, D. A., & Murphy, J. G. (2012). Improved characterization of gas–particle partitioning for per-and polyfluoroalkyl substances in the atmosphere using annular diffusion denuder samplers. *Environmental Science & Technology, 46*(13), 7199–7206.

Ahrens, L., Xie, Z., & Ebinghaus, R. (2010). Distribution of perfluoroalkyl compounds in seawater from Northern Europe, Atlantic Ocean, and Southern Ocean. *Chemosphere, 78*(8), 1011–1016.

Allred, B. M., Lang, J. R., Barlaz, M. A., & Field, J. A. (2015). Physical and biological release of poly-and perfluoroalkyl substances (PFASs) from municipal solid waste in anaerobic model landfill reactors. *Environmental Science & Technology, 49*(13), 7648–7656.

Al-Niami, A. N. S., & Rushton, K. R. (1977). Analysis of flow against dispersion in porous media. *Journal of Hydrology, 33*(1–2), 87–97.

Anderson, R. H., Adamson, D. T., & Stroo, H. F. (2019). Partitioning of poly-and perfluoro-alkyl substances from soil to groundwater within aqueous film-forming foam source zones. *Journal of Contaminant Hydrology, 220*, 59–65.

Anderson, R. H., Long, G. C., Porter, R. C., & Anderson, J. K. (2016). Occurrence of select perfluoroalkyl substances at U.S. Air Force aqueous film-forming foam release sites other than fire-training areas: Field-validation of critical fate and transport properties. *Chemosphere, 150*, 678–685. https://doi.org/10.1016/j.chemosphere.2016.01.014

Anderson, R. H., Feild, J. B., Dieffenbach-Carle, H., Elsharnouby, O., & Krebs, R. K. (2022). Assessment of PFAS in collocated soil and porewater samples at an AFFF-impacted source zone: Field-scale validation of suction lysimeters. *Chemosphere, 308*, 136247.

Anderson, R. H. (2021). The case for direct measures of soil-to-groundwater contaminant mass discharge at AFFF-impacted sites. *Environmental Science & Technology, 55*(10), 6580–6583.

Armitage, J. M., MacLeod, M., & Cousins, I. T. (2009a). Modeling the global fate and transport of perfluorooctanoic acid (PFOA) and perfluorooctanoate (PFO) emitted from direct sources using a multispecies mass balance model. *Environmental Science & Technology, 43*(4), 1134–1140.

Armitage, J. M., Schenker, U., Scheringer, M., Martin, J. W., MacLeod, M., & Cousins, I. T. (2009). Modeling the global fate and transport of perfluorooctane sulfonate (PFOS) and precursor compounds in relation to temporal trends in wildlife exposure. *Environmental Science & Technology, 43*(24), 9274–9280.

Asher, B. J., Wang, Y., De Silva, A. O., Backus, S., Muir, D. C. G., Wong, C. S., & Martin, J. W. (2012). Enantiospecific perfluorooctane sulfonate (PFOS) analysis reveals evidence for the source contribution of PFOS-precursors to the Lake Ontario foodweb. *Environmental Science & Technology, 46*(14), 7653–7660.

ASTM. (2013). *Standard guide for conducting bioconcentration tests with fishes and saltwater bivalve mollusks.* ASTM.

Avendaño, S. M., & Liu, J. (2015). Production of PFOS from aerobic soil biotransformation of two perfluoroalkyl sulfonamide derivatives. *Chemosphere, 119*, 1084–1090.

Baduel, C., Paxman, C. J., & Mueller, J. F. (2015). Perfluoroalkyl substances in a firefighting training ground (FTG), distribution and potential future release. *Journal of Hazardous Materials, 296*, 46–53.

Bai, X., & Son, Y. (2021). Perfluoroalkyl substances (PFAS) in surface water and sediments from two urban watersheds in Nevada, USA. *Science of the Total Environment, 751*, 141622.

Balgooyen, S., & Remucal, C. K. (2022). Tributary loading and sediment desorption as sources of PFAS to receiving waters. *Acs Es&t Water, 2*(3), 436–445.

Barton, C. A., Botelho, M. A., & Kaiser, M. A. (2008). Solid vapor pressure and enthalpy of sublimation for perfluorooctanoic acid. *Journal of Chemical & Engineering Data, 53*(4), 939–941.

Barton, C. A., Butler, L. E., Zarzecki, C. J., Flaherty, J., & Kaiser, M. (2006). Characterizing perfluorooctanoate in ambient air near the fence line of a manufacturing facility: Comparing modeled and monitored values. *Journal of the Air & Waste Management Association, 56*(1), 48–55. https://doi.org/10.1080/10473289.2006.10464429

Barton, C. A., Kaiser, M. A., & Russell, M. H. (2007). Partitioning and removal of perfluorooctanoate during rain events: The importance of physical-chemical properties. *Journal of Environmental Monitoring, 9*(8), 839–846.

Barzen-Hanson, K. A., Davis, S. E., Kleber, M., & Field, J. A. (2017). Sorption of fluorotelomer sulfonates, fluorotelomer sulfonamido betaines, and a fluorotelomer sulfonamido amine in national foam aqueous film-forming foam to soil. *Environmental Science & Technology, 51*(21), 12394–12404.

Benskin, J. P., Muir, D. C. G., Scott, B. F., Spencer, C., De Silva, A. O., Kylin, H., Martin, J. W., Morris, A., Lohmann, R., & Tomy, G. (2012). Perfluoroalkyl acids in the Atlantic and Canadian Arctic oceans. *Environmental Science & Technology*, 46(11), 5815–5823.

Blaine, A. C., Rich, C. D., Hundal, L. S., Lau, C., Mills, M. A., Harris, K. M., & Higgins, C. P. (2013). Uptake of perfluoroalkyl acids into edible crops via land applied biosolids: Field and greenhouse studies. *Environmental Science & Technology*, 47(24), 14062–14069.

Blaine, A. C., Rich, C. D., Sedlacko, E. M., Hundal, L. S., Kumar, K., Lau, C., Mills, M. A., Harris, K. M., & Higgins, C. P. (2014a). Perfluoroalkyl acid distribution in various plant compartments of edible crops grown in biosolids-amended soils. *Environmental Science & Technology*, 48(14), 7858–7865.

Blaine, A. C., Rich, C. D., Sedlacko, E. M., Hyland, K. C., Stushnoff, C., Dickenson, E. R. V, & Higgins, C. P. (2014b). Perfluoroalkyl acid uptake in lettuce (Lactuca sativa) and strawberry (Fragaria ananassa) irrigated with reclaimed water. *Environmental Science & Technology*, 48(24), 14361–14368.

Borthakur, A., Leonard, J., Koutnik, V. S., Ravi, S., & Mohanty, S. K. (2022). Inhalation risks of wind-blown dust from biosolid-applied agricultural lands: Are they enriched with microplastics and PFAS? *Current Opinion in Environmental Science & Health*, 25, 100309.

Borthakur, A., Olsen, P., Dooley, G. P., Cranmer, B. K., Rao, U., Hoek, E. M. V, Blotevogel, J., Mahendra, S., & Mohanty, S. K. (2021). Dry-wet and freeze-thaw cycles enhance PFOA leaching from subsurface soils. *Journal of Hazardous Materials Letters*, 2, 100029.

Borthakur, A., Wang, M., He, M., Ascencio, K., Blotevogel, J., Adamson, D. T., Mahendra, S., & Mohanty, S. K. (2021). Perfluoroalkyl acids on suspended particles: Significant transport pathways in surface runoff, surface waters, and subsurface soils. *Journal of Hazardous Materials*, 417, 126159.

Bossi, R., Vorkamp, K., & Skov, H. (2016). Concentrations of organochlorine pesticides, polybrominated diphenyl ethers and perfluorinated compounds in the atmosphere of North Greenland. *Environmental Pollution*, 217, 4–10. https://doi.org/https://doi.org/10.1016/j.envpol.2015.12.026

Bräunig, J., Baduel, C., Heffernan, A., Rotander, A., Donaldson, E., & Mueller, J. F. (2017). Fate and redistribution of perfluoroalkyl acids through AFFF-impacted groundwater. *Science of the Total Environment*, 596, 360–368.

Brendel, S., Fetter, É., Staude, C., Vierke, L., & Biegel-Engler, A. (2018). Short-chain perfluoroalkyl acids: Environmental concerns and a regulatory strategy under REACH. *Environmental Sciences Europe*, 30, 1–11.

Brusseau, Mark L. (2018). Assessing the potential contributions of additional retention processes to PFAS retardation in the subsurface. *Science of the Total Environment*, 613, 176–185.

Brusseau, Mark L. (2019a). Estimating the relative magnitudes of adsorption to solid-water and air/oil-water interfaces for per-and poly-fluoroalkyl substances. *Environmental Pollution*, 254, 113102.

Brusseau, Mark L. (2019b). The influence of molecular structure on the adsorption of PFAS to fluid-fluid interfaces: Using QSPR to predict interfacial adsorption coefficients. *Water Research*, 152, 148–158.

Brusseau, Mark L. (2020). Simulating PFAS transport influenced by rate-limited multi-process retention. *Water Research*, 168, 115179.

Brusseau, M. L., & Guo, B. (2022). PFAS concentrations in soil versus soil porewater: Mass distributions and the impact of adsorption at air-water interfaces. *Chemosphere*, 302, 134938.

Brusseau, Mark L., Yan, N., Van Glubt, S., Wang, Y., Chen, W., Lyu, Y., Dungan, B., Carroll, K. C., & Holguin, F. O. (2019). Comprehensive retention model for PFAS transport in subsurface systems. *Water Research*, *148*, 41–50.

Buck, R. C., Franklin, J., Berger, U., Conder, J. M., Cousins, I. T., De Voogt, P., Jensen, A. A., Kannan, K., Mabury, S. A., & van Leeuwen, S. P. J. (2011). Perfluoroalkyl and poly-fluoroalkyl substances in the environment: Terminology, classification, and origins. *Integrated Environmental Assessment and Management*, *7*(4), 513–541.

Burkhard, L. P. (2003). Factors influencing the design of bioaccumulation factor and biota-sediment accumulation factor field studies. *Environmental Toxicology and Chemistry: An International Journal*, *22*(2), 351–360.

Burkhard, L. P. (2021). Evaluation of published bioconcentration factor (BCF) and bioac-cumulation factor (BAF) data for per- and polyfluoroalkyl substances across aquatic species. *Environmental Toxicology and Chemistry*, *40*(6), 1530–1543.

Butt, C. M., Berger, U., Bossi, R., & Tomy, G. T. (2010). Levels and trends of poly-and per-fluorinated compounds in the arctic environment. *Science of the Total Environment*, *408*(15), 2936–2965.

Butt, C. M., Mabury, S. A., Kwan, M., Wang, X., & Muir, D. C. G. (2008). Spatial trends of perfluoroalkyl compounds in ringed seals (Phoca hispida) from the Canadian Arctic. *Environmental Toxicology and Chemistry: An International Journal*, *27*(3), 542–553.

Cai, M., Xie, Z., Möller, A., Yin, Z., Huang, P., Cai, M., Yang, H., Sturm, R., He, J., & Ebinghaus, R. (2012b). Polyfluorinated compounds in the atmosphere along a cruise pathway from the Japan Sea to the Arctic Ocean. *Chemosphere*, *87*(9), 989–997.

Cai, M., Yang, H., Xie, Z., Zhao, Z., Wang, F., Lu, Z., Sturm, R., & Ebinghaus, R. (2012c). Per-and polyfluoroalkyl substances in snow, lake, surface runoff water and coastal seawater in Fildes Peninsula, King George Island, Antarctica. *Journal of Hazardous Materials*, *209*, 335–342.

Cai, M., Zhao, Z., Yin, Z., Ahrens, L., Huang, P., Cai, M., Yang, H., He, J., Sturm, R., & Ebinghaus, R. (2012a). Occurrence of perfluoroalkyl compounds in surface waters from the North Pacific to the Arctic Ocean. *Environmental Science & Technology*, *46*(2), 661–668.

Cai, W., Navarro, D. A., Du, J., Ying, G., Yang, B., McLaughlin, M. J., & Kookana, R. S. (2022). Increasing ionic strength and valency of cations enhance sorption through hydrophobic interactions of PFAS with soil surfaces. *Science of the Total Environment*, *817*, 152975.

Campo, J., Lorenzo, M., Pérez, F., Picó, Y., la Farré, M., & Barceló, D. (2016). Analysis of the presence of perfluoroalkyl substances in water, sediment and biota of the Jucar River (E Spain). Sources, partitioning and relationships with water physical characteristics. *Environmental Research*, *147*, 503–512.

Campos-Pereira, H., Kleja, D. B., Sjöstedt, C., Ahrens, L., Klysubun, W., & Gustafsson, J. P. (2020). The adsorption of per-and polyfluoroalkyl substances (PFASs) onto fer-rihydrite is governed by surface charge. *Environmental Science & Technology*, *54*(24), 15722–15730.

Casas, G., Martínez-Varela, A., Roscales, J. L., Vila-Costa, M., Dachs, J., & Jiménez, B. (2020). Enrichment of perfluoroalkyl substances in the sea-surface microlayer and sea-spray aerosols in the Southern Ocean. *Environmental Pollution*, *267*, 115512.

Chen, H., Reinhard, M., Nguyen, V. T., & Gin, K. Y.-H. (2016a). Reversible and irrevers-ible sorption of perfluorinated compounds (PFCs) by sediments of an urban reservoir. *Chemosphere*, *144*, 1747–1753.

Chen, S., Jiao, X.-C., Gai, N., Li, X.-J., Wang, X.-C., Lu, G.-H., Piao, H.-T., Rao, Z., & Yang, Y.-L. (2016b). Perfluorinated compounds in soil, surface water, and groundwater from rural areas in eastern China. *Environmental Pollution*, *211*, 124–131.

Choi, Y. J., Helbling, D. E., Liu, J., Olivares, C. I., & Higgins, C. P. (2022). Microbial biotransformation of aqueous film-forming foam derived polyfluoroalkyl substances. *Science of the Total Environment*, *824*, 153711.

Codling, G., Halsall, C., Ahrens, L., Del Vento, S., Wiberg, K., Bergknut, M., Laudon, H., & Ebinghaus, R. (2014). The fate of per-and polyfluoroalkyl substances within a melting snowpack of a boreal forest. *Environmental Pollution*, *191*, 190–198.

Codling, G., Sturchio, N. C., Rockne, K. J., Li, A., Peng, H., Timothy, J. T., Jones, P. D., & Giesy, J. P. (2018). Spatial and temporal trends in poly-and per-fluorinated compounds in the Laurentian Great Lakes Erie, Ontario and St. Clair. *Environmental Pollution*, *237*, 396–405.

Conder, J. M., Hoke, R. A., Wolf, W. de, Russell, M. H., & Buck, R. C. (2008). Are PFCAs bioaccumulative? A critical review and comparison with regulatory criteria and persistent lipophilic compounds. *Environmental Science & Technology*, *42*(4), 995–1003.

Conder, J., Arblaster, J., Larson, E., Brown, J., & Higgins, C. (2020). Guidance for assessing the ecological risks of PFASs to threatened and endangered species at aqueous film forming foam-impacted sites. *SERDP Contract Report ER18-1614*. SERDP.

Costanza, J., Abriola, L. M., & Pennell, K. D. (2020). Aqueous film-forming foams exhibit greater interfacial activity than PFOA, PFOS, or FOSA. *Environmental Science & Technology*, *54*(21), 13590–13597.

Costanza, J., Arshadi, M., Abriola, L. M., & Pennell, K. D. (2019). Accumulation of PFOA and PFOS at the air–water interface. *Environmental Science & Technology Letters*, *6*(8), 487–491.

D'Agostino, L. A., & Mabury, S. A. (2017). Aerobic biodegradation of 2 fluorotelomer sulfonamide–based aqueous film–forming foam components produces perfluoroalkyl carboxylates. *Environmental Toxicology and Chemistry*, *36*(8), 2012–2021.

D'Ambro, E. L., Pye, H. O. T., Bash, J. O., Bowyer, J., Allen, C., Efstathiou, C., Gilliam, R. C., Reynolds, L., Talgo, K., & Murphy, B. N. (2021). Characterizing the air emissions, transport, and deposition of per-and polyfluoroalkyl substances from a fluoropolymer manufacturing facility. *Environmental Science & Technology*, *55*(2), 862–870.

D'eon, J. C., Hurley, M. D., Wallington, T. J., & Mabury, S. A. (2006). Atmospheric chemistry of N-methyl perfluorobutane sulfonamidoethanol, $C_4F_9SO_2N$ (CH_3) CH_2CH_2OH: Kinetics and mechanism of reaction with OH. *Environmental Science & Technology*, *40*(6), 1862–1868.

da Silva, B. F., Aristizabal-Henao, J. J., Aufmuth, J., Awkerman, J., & Bowden, J. A. (2022). Survey of per-and polyfluoroalkyl substances (PFAS) in surface water collected in Pensacola, FL. *Heliyon*, *8*(8), e10182.

Dassuncao, C., Hu, X. C., Zhang, X., Bossi, R., Dam, M., Mikkelsen, B., & Sunderland, E. M. (2017). Temporal shifts in poly-and perfluoroalkyl substances (PFASs) in North Atlantic pilot whales indicate large contribution of atmospheric precursors. *Environmental Science & Technology*, *51*(8), 4512–4521.

Dassuncao, C., Pickard, H., Pfohl, M., Tokranov, A. K., Li, M., Mikkelsen, B., Slitt, A., & Sunderland, E. M. (2019). Phospholipid levels predict the tissue distribution of poly-and perfluoroalkyl substances in a marine mammal. *Environmental Science & Technology Letters*, *6*(3), 119–125.

Dasu, K., & Lee, L. S. (2016). Aerobic biodegradation of toluene-2, 4-di (8: 2 fluorotelomer urethane) and hexamethylene-1, 6-di (8: 2 fluorotelomer urethane) monomers in soils. *Chemosphere*, *144*, 2482–2488.

Dasu, K., Liu, J., & Lee, L. S. (2012). Aerobic soil biodegradation of 8: 2 fluorotelomer stearate monoester. *Environmental Science & Technology*, *46*(7), 3831–3836.

Davis, K. L., Aucoin, M. D., Larsen, B. S., Kaiser, M. A., & Hartten, A. S. (2007). Transport of ammonium perfluorooctanoate in environmental media near a fluoropolymer manufacturing facility. *Chemosphere*, *67*(10), 2011–2019.

Dawson, D. E., Lau, C., Pradeep, P., Sayre, R. R., Judson, R. S., Tornero-Velez, R., & Wambaugh, J. F. (2023). A machine learning model to estimate toxicokinetic half-lives of per-and polyfluoro-alkyl substances (PFAS) in multiple species. *Toxics*, *11*(2), 98. https://doi.org/10.3390/toxics11020098

De Silva, A. O., Armitage, J. M., Bruton, T. A., Dassuncao, C., Heiger-Bernays, W., Hu, X. C., Kärrman, A., Kelly, B., Ng, C., & Robuck, A. (2021). PFAS exposure pathways for humans and wildlife: A synthesis of current knowledge and key gaps in understanding. *Environmental Toxicology and Chemistry*, *40*(3), 631–657.

De Silva, A. O., Muir, D. C. G., & Mabury, S. A. (2009). Distribution of perfluorocarboxylate isomers in select samples from the North American environment. *Environmental Toxicology and Chemistry: An International Journal*, *28*(9), 1801–1814.

Deng, S., Zhang, Q., Nie, Y., Wei, H., Wang, B., Huang, J., Yu, G., & Xing, B. (2012). Sorption mechanisms of perfluorinated compounds on carbon nanotubes. *Environmental Pollution*, *168*, 138–144.

Dinglasan, M. J. A., Ye, Y., Edwards, E. A., & Mabury, S. A. (2004). Fluorotelomer alcohol biodegradation yields poly-and perfluorinated acids. *Environmental Science & Technology*, *38*(10), 2857–2864.

DiStefano, R., Feliciano, T., Mimna, R. A., Redding, A. M., & Matthis, J. (2022). Thermal destruction of PFAS during full-scale reactivation of PFAS-laden granular activated carbon. *Remediation Journal*, *32*(4), 231–238. https://doi.org/https://doi.org/10.1002/rem.21735

Downer, A., Eastoe, J., Pitt, A. R., Penfold, J., & Heenan, R. K. (1999). Adsorption and micellisation of partially-and fully-fluorinated surfactants. *Colloids and Surfaces A: Physicochemical and Engineering Aspects*, *156*(1–3), 33–48.

Dreyer, A, Kirchgeorg, T., Weinberg, I., & Matthias, V. (2015). Particle-size distribution of airborne poly-and perfluorinated alkyl substances. *Chemosphere*, *129*, 142–149.

Dreyer, A., Matthias, V., Weinberg, I., & Ebinghaus, R. (2010). Wet deposition of poly-and perfluorinated compounds in Northern Germany. *Environmental Pollution*, *158*(5), 1221–1227.

Dreyer, Annekatrin, Weinberg, I., Temme, C., & Ebinghaus, R. (2009). Polyfluorinated compounds in the atmosphere of the Atlantic and Southern Oceans: Evidence for a global distribution. *Environmental Science & Technology*, *43*(17), 6507–6514. https://doi.org/10.1021/es9010465

Du, Z., Deng, S., Bei, Y., Huang, Q., Wang, B., Huang, J., & Yu, G. (2014). Adsorption behavior and mechanism of perfluorinated compounds on various adsorbents—A review. *Journal of Hazardous Materials*, *274*, 443–454.

Ebrahimi, F., Lewis, A. J., Sales, C. M., Suri, R., & McKenzie, E. R. (2021). Linking PFAS partitioning behavior in sewage solids to the solid characteristics, solution chemistry, and treatment processes. *Chemosphere*, *271*, 129530.

Ellis, D. A., Martin, J. W., De Silva, A. O., Mabury, S. A., Hurley, M. D., Sulbaek Andersen, M. P., & Wallington, T. J. (2004). Degradation of fluorotelomer alcohols: A likely atmospheric source of perfluorinated carboxylic acids. *Environmental Science & Technology*, *38*(12), 3316–3321.

Filipovic, M., Laudon, H., McLachlan, M. S., & Berger, U. (2015). Mass balance of perfluorinated alkyl acids in a pristine boreal catchment. *Environmental Science & Technology*, *49*(20), 12127–12135.

Fitzgerald, N. J. M., Wargenau, A., Sorenson, C., Pedersen, J., Tufenkji, N., Novak, P. J., & Simcik, M. F. (2018). Partitioning and accumulation of perfluoroalkyl substances in model lipid bilayers and bacteria. *Environmental Science & Technology, 52*(18), 10433–10440.

Franklin, J. (2016). How reliable are field-derived biomagnification factors and trophic magnification factors as indicators of bioaccumulation potential? Conclusions from a case study on per- and polyfluoroalkyl substances. *Integrated Environmental Assessment and Management, 12*(1), 6–20.

Frisbee, S. J., Brooks, A. P., Maher, A., Flensborg, P., Arnold, S., Fletcher, T., Steenland, K., Shankar, A., Knox, S. S., Pollard, C., Halverson, J. A., Vieira, V. M., Jin, C., Leyden, K. M., & Ducatman, A. M. (2009). The C8 health project: Design, methods, and participants. *Environmental Health Perspectives, 117*(12), 1873–1882. https://doi.org/10.1289/ehp.0800379

Galatius, A., Bossi, R., Sonne, C., Rigét, F. F., Kinze, C. C., Lockyer, C., Teilmann, J., & Dietz, R. (2013). PFAS profiles in three North Sea top predators: Metabolic differences among species? *Environmental Science and Pollution Research, 20*, 8013–8020.

Galloway, J. E., Moreno, A. V. P., Lindstrom, A. B., Strynar, M. J., Newton, S., May, A. A., & Weavers, L. K. (2020). Evidence of air dispersion: HFPO–DA and PFOA in Ohio and West Virginia surface water and soil near a fluoropolymer production facility. *Environmental Science & Technology, 54*(12), 7175–7184.

Garg, A., Shetti, N. P., Basu, S., Nadagouda, M. N., Tejraj, M., Aminabhavi, T. N. (2023). Treatment technologies for removal of per- and polyfluoroalkyl substances (PFAS) in biosolids. *Chemical Engineering Journal, 453*, 139964.

Gauthier, S. A., & Mabury, S. A. (2005). Aqueous photolysis of 8: 2 fluorotelomer alcohol. *Environmental Toxicology and Chemistry: An International Journal, 24*(8), 1837–1846.

Gaylard, S. (2017). *Per and polyfluorinated alkyl substances (PFAS) in the marine environment–Preliminary ecological findings.* Environmental Protection Authority.

Ge, H., Yamazaki, E., Yamashita, N., Taniyasu, S., Ogata, A., & Furuuchi, M. (2017). Particle size specific distribution of perfluoro alkyl substances in atmospheric particulate matter in Asian cities. *Environmental Science: Processes & Impacts, 19*(4), 549–560.

Gebbink, W. A., Berger, U., & Cousins, I. T. (2015). Estimating human exposure to PFOS isomers and PFCA homologues: The relative importance of direct and indirect (precursor) exposure. *Environment International, 74*, 160–169.

Gellrich, V., Stahl, T., & Knepper, T. P. (2012). Behavior of perfluorinated compounds in soils during leaching experiments. *Chemosphere, 87*(9), 1052–1056.

Ghisi, R., Vamerali, T., & Manzetti, S. (2019). Accumulation of perfluorinated alkyl substances (PFAS) in agricultural plants: A review. *Environmental Research, 169*, 326–341.

Glüge, J., Scheringer, M., Cousins, I. T., DeWitt, J. C., Goldenman, G., Herzke, D., Lohmann, R., Ng, C. A., Trier, X., & Wang, Z. (2020). An overview of the uses of per-and polyfluoroalkyl substances (PFAS). *Environmental Science: Processes & Impacts, 22*(12), 2345–2373. https://doi.org/10.1039/D0EM00291G

Gobas, F. A. P. C., Lee, Y., Lo, J. C., Parkerton, T. F., & Letinski, D. J. (2020b). A toxicokinetic framework and analysis tool for interpreting Organisation for Economic Co-Operation and Development guideline 305 dietary bioaccumulation tests. *Environmental Toxicology and Chemistry, 39*(1), 171–188.

Gobas, F., Kelly, B., & Kim, J. (2020a). *A framework for assessing bioaccumulation and exposure risks of per and polyfluoroalkyl substances in threatened and endangered species on aqueous film forming foam (AFFF)-impacted sites.* Strategic Environmental Research and Development Program.

Gobelius, L., Lewis, J., & Ahrens, L. (2017). Plant uptake of per-and polyfluoroalkyl substances at a contaminated fire training facility to evaluate the phytoremediation potential of various plant species. *Environmental Science & Technology, 51*(21), 12602–12610.

Goeritz, I., Falk, S., Stahl, T., Schäfers, C., & Schlechtriem, C. (2013). Biomagnification and tissue distribution of perfluoroalkyl substances (PFASs) in market-size rainbow trout (Oncorhynchus mykiss). *Environmental Toxicology and Chemistry, 32*(9), 2078–2088.

Guelfo, J. L., & Higgins, C. P. (2013). Subsurface transport potential of perfluoroalkyl acids at aqueous film-forming foam (AFFF)-impacted sites. *Environmental Science & Technology, 47*(9), 4164–4171.

Guelfo, J. L., Wunsch, A., McCray, J., Stults, J. F., & Higgins, C. P. (2020). Subsurface transport potential of perfluoroalkyl acids (PFAAs): Column experiments and modeling. *Journal of Contaminant Hydrology, 233*, 103661.

Guo, B., Zeng, J., & Brusseau, M. L. (2020). A mathematical model for the release, transport, and retention of per- and polyfluoroalkyl substances (PFAS) in the vadose zone. *Water Resources Research, 56*(2), e2019WR026667.

Han, X., Snow, T. A., Kemper, R. A., & Jepson, G. W. (2003). Binding of perfluorooctanoic acid to rat and human plasma proteins. *Chemical Research in Toxicology, 16*(6), 775–781.

Harding-Marjanovic, K. C., Houtz, E. F., Yi, S., Field, J. A., Sedlak, D. L., & Alvarez-Cohen, L. (2015). Aerobic biotransformation of fluorotelomer thioether amido sulfonate (Lodyne) in AFFF-amended microcosms. *Environmental Science & Technology, 49*(13), 7666–7674.

Hellsing, M. S., Josefsson, S., Hughes, A. V., & Ahrens, L. (2016). Sorption of perfluoroalkyl substances to two types of minerals. *Chemosphere, 159*, 385–391.

Henry, B. J., Carlin, J. P., Hammerschmidt, J. A., Buck, R. C., Buxton, L. W., Fiedler, H., Seed, J., & Hernandez, O. (2018). A critical review of the application of polymer of low concern and regulatory criteria to fluoropolymers. *Integrated Environmental Assessment and Management, 14*(3), 316–334.

Higgins, C. P., & Luthy, R. G. (2006). Sorption of perfluorinated surfactants on sediments. *Environmental Science & Technology, 40*(23), 7251–7256.

Horst, J., McDonough, J., Ross, I., Dickson, M., Miles, J., Hurst, J., & Storch, P. (2018). Water treatment technologies for PFAS: The next generation. *Groundwater Monitoring & Remediation, 38*(2), 13–23.

Houde, M., Bujas, T. A. D., Small, J., Wells, R. S., Fair, P. A., Bossart, G. D., Solomon, K. R., & Muir, D. C. G. (2006). Biomagnification of perfluoroalkyl compounds in the bottlenose dolphin (Tursiops truncatus) food web. *Environmental Science & Technology, 40*(13), 4138–4144.

Houde, M., Czub, G., Small, J. M., Backus, S., Wang, X., Alaee, M., & Muir, D. C. G. (2008). Fractionation and bioaccumulation of perfluorooctane sulfonate (PFOS) isomers in a Lake Ontario food web. *Environmental Science & Technology, 42*(24), 9397–9403.

Houde, M., De Silva, A. O., Muir, D. C. G., & Letcher, R. J. (2011). Monitoring of perfluorinated compounds in aquatic biota: An updated review: PFCs in aquatic biota. *Environmental Science & Technology, 45*(19), 7962–7973.

Houtz, E. F, Higgins, C. P., Field, J. A., & Sedlak, D. L. (2013). Persistence of perfluoroalkyl acid precursors in AFFF-impacted groundwater and soil. *Environmental Science & Technology, 47*(15), 8187–8195.

Houtz, E. F., & Sedlak, D. L. (2012). Oxidative conversion as a means of detecting precursors to perfluoroalkyl acids in urban runoff. *Environmental Science & Technology, 46*(17), 9342–9349.

Hu, X. C., Ge, B., Ruyle, B. J., Sun, J., & Sunderland, E. M. (2021). A statistical approach for identifying private wells susceptible to perfluoroalkyl substances (PFAS) contamination. *Environmental Science & Technology Letters*, *8*(7), 596–602. https://doi.org/10.1021/acs.estlett.1c00264

Huang, S., & Jaffé, P. R. (2019). Defluorination of perfluorooctanoic acid (PFOA) and perfluorooctane sulfonate (PFOS) by Acidimicrobium sp. strain A6. *Environmental Science & Technology*, *53*(19), 11410–11419.

Hurley, M. D., Sulbaek Andersen, M. P., Wallington, T. J., Ellis, D. A., Martin, J. W., & Mabury, S. A. (2004). Atmospheric chemistry of perfluorinated carboxylic acids: Reaction with OH radicals and atmospheric lifetimes. *The Journal of Physical Chemistry A*, *108*(4), 615–620. https://doi.org/10.1021/jp036343b

IARC. (2016). *Some chemicals used as solvents and in polymer manufacture*. World Health Organization Lyon.

Jin, H., Shan, G., Zhu, L., Sun, H., & Luo, Y. (2018). Perfluoroalkyl acids including isomers in tree barks from a Chinese fluorochemical manufacturing park: Implication for airborne transportation. *Environmental Science & Technology*, *52*(4), 2016–2024.

Jing, P., Rodgers, P. J., & Amemiya, S. (2009). High lipophilicity of perfluoroalkyl carboxylate and sulfonate: Implications for their membrane permeability. *Journal of the American Chemical Society*, *131*(6), 2290–2296.

Joerss, H., Xie, Z., Wagner, C. C., Von Appen, W.-J., Sunderland, E. M., & Ebinghaus, R. (2020). Transport of legacy perfluoroalkyl substances and the replacement compound HFPO-DA through the Atlantic Gateway to the Arctic Ocean—is the Arctic a sink or a source? *Environmental Science & Technology*, *54*(16), 9958–9967.

Johnson, R. L., Anschutz, A. J., Smolen, J. M., Simcik, M. F., & Penn, R. L. (2007). The adsorption of perfluorooctane sulfonate onto sand, clay, and iron oxide surfaces. *Journal of Chemical & Engineering Data*, *52*(4), 1165–1170.

Jones, P. D., Hu, W., De Coen, W., Newsted, J. L., & Giesy, J. P. (2003). Binding of perfluorinated fatty acids to serum proteins. *Environmental Toxicology and Chemistry: An International Journal*, *22*(11), 2639–2649.

Kaiser, M. A., Dawson, B. J., Barton, C. A., & Botelho, M. A. (2010). Understanding potential exposure sources of perfluorinated carboxylic acids in the workplace. *Annals of Occupational Hygiene*, *54*(8), 915–922.

Kaiser, M. A., Larsen, B. S., Kao, C.-P. C., & Buck, R. C. (2005). Vapor pressures of perfluorooctanoic,-nonanoic,-decanoic,-undecanoic, and-dodecanoic acids. *Journal of Chemical & Engineering Data*, *50*(6), 1841–1843.

Kancharla, S., Choudhary, A., Davis, R. T., Dong, D., Bedrov, D., Tsianou, M., & Alexandridis, P. (2022). GenX in water: Interactions and self-assembly. *Journal of Hazardous Materials*, *428*, 128137.

Kim, S.-K., & Kannan, K. (2007). Perfluorinated acids in air, rain, snow, surface runoff, and lakes: Relative importance of pathways to contamination of Urban Lakes. *Environmental Science & Technology*, *41*(24), 8328–8334. https://doi.org/10.1021/es072107t

Kirchgeorg, T, Dreyer, A., Gabrielli, P., Gabrieli, J., Thompson, L. G., Barbante, C., & Ebinghaus, R. (2016). Seasonal accumulation of persistent organic pollutants on a high altitude glacier in the Eastern Alps. *Environmental Pollution*, *218*, 804–812.

Kirchgeorg, Torben, Dreyer, A., Gabrieli, J., Kehrwald, N., Sigl, M., Schwikowski, M., Boutron, C., Gambaro, A., Barbante, C., & Ebinghaus, R. (2013). Temporal variations of perfluoroalkyl substances and polybrominated diphenyl ethers in alpine snow. *Environmental Pollution*, *178*, 367–374.

Knight, E. R., Janik, L. J., Navarro, D. A., Kookana, R. S., & McLaughlin, M. J. (2019). Predicting partitioning of radiolabelled 14C-PFOA in a range of soils using diffuse reflectance infrared spectroscopy. *Science of the Total Environment*, *686*, 505–513.

Krafft, M. P., & Riess, J. G. (2015). Selected physicochemical aspects of poly-and perfluoroalkylated substances relevant to performance, environment and sustainability—Part one. *Chemosphere*, *129*, 4–19.

Krippner, J., Falk, S., Brunn, H., Georgii, S., Schubert, S., & Stahl, T. (2015). Accumulation potentials of perfluoroalkyl carboxylic acids (PFCAs) and perfluoroalkyl sulfonic acids (PFSAs) in maize (Zea mays). *Journal of Agricultural and Food Chemistry*, *63*(14), 3646–3653.

Krug, J. D., Lemieux, P. M., Lee, C.-W., Ryan, J. V, Kariher, P. H., Shields, E. P., Wickersham, L. C., Denison, M. K., Davis, K. A., Swensen, D. A., Burnette, R. P., Wendt, J. O. L., & Linak, W. P. (2022). Combustion of C1 and C2 PFAS: Kinetic modeling and experiments. *Journal of the Air & Waste Management Association*, *72*(3), 256–270. https://doi.org/10.1080/10962247.2021.2021317

Kwok, K. Y., Taniyasu, S., Yeung, L. W. Y., Murphy, M. B., Lam, P. K. S., Horii, Y., Kannan, K., Petrick, G., Sinha, R. K., & Yamashita, N. (2010). Flux of perfluorinated chemicals through wet deposition in Japan, the United States, and several other countries. *Environmental Science & Technology*, *44*(18), 7043–7049. https://doi.org/10.1021/es101170c

Lai, S., Song, J., Song, T., Huang, Z., Zhang, Y., Zhao, Y., Liu, G., Zheng, J., Mi, W., Tang, J., Zou, S., Ebinghaus, R., & Xie, Z. (2016). Neutral polyfluoroalkyl substances in the atmosphere over the northern South China Sea. *Environmental Pollution*, *214*, 449–455. https://doi.org/https://doi.org/10.1016/j.envpol.2016.04.047

Lampert, D. J. (2018). Emerging research needs for assessment and remediation of sediments contaminated with per-and poly-fluoroalkyl substances. *Current Pollution Reports*, *4*, 277–279.

Langberg, H. A., Breedveld, G. D., Slinde, G. A., Grønning, H. M., Høisæter, Å., Jartun, M., Rundberget, T., Jenssen, B. M., & Hale, S. E. (2020). Fluorinated precursor compounds in sediments as a source of perfluorinated alkyl acids (PFAA) to biota. *Environmental Science & Technology*, *54*(20), 13077–13089.

Langberg, H. A., Hale, S. E., Breedveld, G. D., Jenssen, B. M., & Jartun, M. (2022). A review of PFAS fingerprints in fish from Norwegian freshwater bodies subject to different source inputs. *Environmental Science: Processes & Impacts*, *24*(2), 330–342.

Le, S.-T., Gao, Y., Kibbey, T. C. G., Glamore, W. C., & O'Carroll, D. M. (2021). A new framework for modeling the effect of salt on interfacial adsorption of PFAS in environmental systems. *Science of The Total Environment*, *796*, 148893.

Lechner, M., & Knapp, H. (2011). Carryover of perfluorooctanoic acid (PFOA) and perfluorooctane sulfonate (PFOS) from soil to plant and distribution to the different plant compartments studied in cultures of carrots (Daucus carota ssp. Sativus), potatoes (Solanum tuberosum), and cucumber. *Journal of Agricultural and Food Chemistry*, *59*(20), 11011–11018.

Lee, H., Tevlin, A. G., Mabury, S. A., & Mabury, S. A. (2014). Fate of polyfluoroalkyl phosphate diesters and their metabolites in biosolids-applied soil: Biodegradation and plant uptake in greenhouse and field experiments. *Environmental Science & Technology*, *48*(1), 340–349.

Leidos. (2019). *Final surface water study technical memorandum for the perfluorinated compound facility investigation at Horsham Air Guard Station (111th Attack Wing)*. U.S. Air National Guard.

Lescord, G. L., Kidd, K. A., De Silva, A. O., Williamson, M., Spencer, C., Wang, X., & Muir, D. C. G. (2015). Perfluorinated and polyfluorinated compounds in lake food webs from the Canadian High Arctic. *Environmental Science & Technology, 49*(5), 2694–2702.

Lesmeister, L., Lange, F. T., Breuer, J., Biegel-Engler, A., Giese, E., & Scheurer, M. (2021). Extending the knowledge about PFAS bioaccumulation factors for agricultural plants– A review. *Science of The Total Environment, 766*, 142640.

Li, Y., Oliver, D. P., & Kookana, R. S. (2018). A critical analysis of published data to discern the role of soil and sediment properties in determining sorption of per and polyfluoro-alkyl substances (PFASs). *Science of the Total Environment, 628*, 110–120.

Lin, A. Y.-C., Panchangam, S. C., Tsai, Y.-T., & Yu, T.-H. (2014). Occurrence of perfluorinated compounds in the aquatic environment as found in science park effluent, river water, rainwater, sediments, and biotissues. *Environmental Monitoring and Assessment, 186*(5), 3265–3275. https://doi.org/10.1007/s10661-014-3617-9

Lin, Y., Capozzi, S. L., Lin, L., & Rodenburg, L. A. (2021). Source apportionment of perfluoro-roalkyl substances in Great Lakes fish. *Environmental Pollution, 290*, 118047.

Lindstrom, A. B., Strynar, M. J., Delinsky, A. D., Nakayama, S. F., McMillan, L., Libelo, E. L., Neill, M., & Thomas, L. (2011). Application of WWTP biosolids and resulting per-fluorinated compound contamination of surface and well water in Decatur, Alabama, USA. *Environmental Science & Technology, 45*(19), 8015–8021.

Liu, J., Lee, L. S., Nies, L. F., Nakatsu, C. H., & Turco, R. F. (2007). Biotransformation of 8: 2 fluorotelomer alcohol in soil and by soil bacteria isolates. *Environmental Science & Technology, 41*(23), 8024–8030.

Liu, W., Jin, Y., Quan, X., Sasaki, K., Saito, N., Nakayama, S. F., Sato, I., & Tsuda, S. (2009). Perfluorosulfonates and perfluorocarboxylates in snow and rain in Dalian, China. *Environment International, 35*(4), 737–742. https://doi.org/https://doi.org/10.1016/j.envint.2009.01.016

Liu, Z., Lu, Y., Song, X., Jones, K., Sweetman, A. J., Johnson, A. C., Zhang, M., Lu, X., & Su, C. (2019). Multiple crop bioaccumulation and human exposure of perfluoroalkyl sub-stances around a mega fluorochemical industrial park, China: Implication for planting optimization and food safety. *Environment International, 127*, 671–684.

Löfstedt Gilljam, J., Leonel, J., Cousins, I. T., & Benskin, J. P. (2016). Is ongoing sulflura-mid use in South America a significant source of perfluorooctanesulfonate (PFOS)? Production inventories, environmental fate, and local occurrence. *Environmental Science & Technology, 50*(2), 653–659.

Lyu, Y., & Brusseau, M. L. (2020). The influence of solution chemistry on air-water inter-facial adsorption and transport of PFOA in unsaturated porous media. *Science of the Total Environment, 713*, 136744.

Martin, J. W., Asher, B. J., Beesoon, S., Benskin, J. P., & Ross, M. S. (2010). PFOS or PreFOS? Are perfluorooctane sulfonate precursors (PreFOS) important determinants of human and environmental perfluorooctane sulfonate (PFOS) exposure? *Journal of Environmental Monitoring, 12*(11), 1979–2004.

Martin, J. W., Ellis, D. A., Mabury, S. A., Hurley, M. D., & Wallington, T. J. (2006). Atmospheric chemistry of perfluoroalkanesulfonamides: Kinetic and product studies of the OH radical and Cl atom initiated oxidation of N-ethyl perfluorobutanesulfon-amide. *Environmental Science & Technology, 40*(3), 864–872.

Martin, J. W., Mabury, S. A., Solomon, K. R., & Muir, D. C. G. (2003). Bioconcentration and tissue distribution of perfluorinated acids in rainbow trout (Oncorhynchus mykiss). *Environmental Toxicology and Chemistry: An International Journal, 22*(1), 196–204.

Martin, J. W., Mabury, S. A., Solomon, K. R., & Muir, D. C. G. (2013). Progress toward under-standing the bioaccumulation of perfluorinated alkyl acids. *Environmental Toxicology and Chemistry, 32*(11), 2421–2423.

Martin, J. W., Muir, D. C. G., Moody, C. A., Ellis, D. A., Kwan, W. C., Solomon, K. R., & Mabury, S. A. (2002). Collection of airborne fluorinated organics and analysis by gas chromatography/chemical ionization mass spectrometry. *Analytical Chemistry, 74*(3), 584–590.

Martin, J. W., Whittle, D. M., Muir, D. C. G., & Mabury, S. A. (2004). Perfluoroalkyl contaminants in a food web from Lake Ontario. *Environmental Science & Technology, 38*(20), 5379–5385.

McGuire, M. E., Schaefer, C., Richards, T., Backe, W. J., Field, J. A., Houtz, E., Sedlak, D. L., Guelfo, J. L., Wunsch, A., & Higgins, C. P. (2014). Evidence of remediation-induced alteration of subsurface poly- and perfluoroalkyl substance distribution at a former firefighter training area. *Environmental Science & Technology, 48*(12), 6644–6652. https://doi.org/10.1021/es5006187

McKenzie, E. R., Siegrist, R. L., McCray, J. E., & Higgins, C. P. (2015). Effects of chemical oxidants on perfluoroalkyl acid transport in one-dimensional porous media columns. *Environmental Science & Technology, 49*(3), 1681–1689.

McKenzie, E. R., Siegrist, R. L., McCray, J. E., & Higgins, C. P. (2016). The influence of a non-aqueous phase liquid (NAPL) and chemical oxidant application on perfluoroalkyl acid (PFAA) fate and transport. *Water Research, 92*, 199–207.

McMurdo, C. J., Ellis, D. A., Webster, E., Butler, J., Christensen, R. D., & Reid, L. K. (2008). Aerosol enrichment of the surfactant PFO and mediation of the water–air transport of gaseous PFOA. *Environmental Science & Technology, 42*(11), 3969–3974.

Mejia-Avendaño, S., Vo Duy, S., Sauvé, S., & Liu, J. (2016). Generation of perfluoroalkyl acids from aerobic biotransformation of quaternary ammonium polyfluoroalkyl surfactants. *Environmental Science & Technology, 50*(18), 9923–9932.

Mejia-Avendaño, S., Zhi, Y., Yan, B., & Liu, J. (2020). Sorption of polyfluoroalkyl surfactants on surface soils: Effect of molecular structures, soil properties, and solution chemistry. *Environmental Science & Technology, 54*(3), 1513–1521.

Mudumbi, J. B. N., Ntwampe, S. K. O., Muganza, M., & Okonkwo, J. O. (2014). Susceptibility of riparian wetland plants to perfluorooctanoic acid (PFOA) accumulation. *International Journal of Phytoremediation, 16*(9), 926–936.

Mussabek, D., Ahrens, L., Persson, K. M., & Berndtsson, R. (2019). Temporal trends and sediment–water partitioning of per-and polyfluoroalkyl substances (PFAS) in lake sediment. *Chemosphere, 227*, 624–629.

Mussabek, D., Persson, K. M., Berndtsson, R., Ahrens, L., Nakagawa, K., & Imura, T. (2020). Impact of the sediment organic vs. mineral content on distribution of the per-and polyfluoroalkyl substances (PFAS) in lake sediment. *International Journal of Environmental Research and Public Health, 17*(16), 5642.

Newell, C. J., Adamson, D. T., Kulkarni, P. R., Nzeribe, B. N., Connor, J. A., Popovic, J., & Stroo, H. F. (2021a). Monitored natural attenuation to manage PFAS impacts to groundwater: Scientific basis. *Groundwater Monitoring & Remediation, 41*(4), 76–89.

Newell, C. J., Adamson, D. T., Kulkarni, P. R., Nzeribe, B. N., Connor, J. A., Popovic, J., & Stroo, H. F. (2021b). Monitored natural attenuation to manage PFAS impacts to groundwater: Potential guidelines. *Remediation Journal, 31*(4), 7–17.

Ney, R. E. (1995). *Fate and transport of organic chemicals in the environment: A practical guide* (2nd ed.). CRC Press.

Ng, C. A., & Hungerbühler, K. (2014). Bioaccumulation of perfluorinated alkyl acids: Observations and models. *Environmental Science & Technology, 48*(9), 4637–4648.

Ng, C. A., & Hungerbuehler, K. (2015). Exploring the use of molecular docking to identify bioaccumulative perfluorinated alkyl acids (PFAAs). *Environmental Science & Technology, 49*(20), 12306–12314.

Nguyen, T. M. H., Bräunig, J., Thompson, K., Thompson, J., Kabiri, S., Navarro, D. A., Kookana, R. S., Grimison, C., Barnes, C. M., & Higgins, C. P. (2020). Influences of chemical properties, soil properties, and solution pH on soil–water partitioning coefficients of per-and polyfluoroalkyl substances (PFASs). *Environmental Science & Technology, 54*(24), 15883–15892.

Nickerson, A., Maizel, A. C., Olivares, C. I., Schaefer, C. E., & Higgins, C. P. (2021). Simulating impacts of biosparging on release and transformation of poly-and perfluorinated alkyl substances from aqueous film-forming foam-impacted soil. *Environmental Science & Technology, 55*(23), 15744–15753.

NYSDOH. (2016). *Village of Hoosick Falls and Town of Hoosick private well sampling, perfluorooctanoic acid (PFOA) results map.* chrome-extension://efaidnbmnnnibpca jpcglclefindmkaj/https://www.villageofhoosickfalls.com/Water/Documents/NYSDOH -WellResults-Hoosick-03112016.pdf

OECD. (2012). *OECD Guidelines for testing of chemicals: Bioaccumulation in fish: Aqueous and dietary exposure.* OECD Publishing.

Pedone, L., Chillura Martino, D., Caponetti, E., Floriano, M. A., & Triolo, R. (1997). Determination of the composition of mixed hydrogenated and fluorinated micelles by small angle neutron scattering. *The Journal of Physical Chemistry B, 101*(46), 9525–9531.

Pereira, H. C., Ullberg, M., Kleja, D. B., Gustafsson, J. P., & Ahrens, L. (2018). Sorption of perfluoroalkyl substances (PFASs) to an organic soil horizon–Effect of cation composition and pH. *Chemosphere, 207*, 183–191.

Pétré, M.-A., Genereux, D. P., Koropeckyj-Cox, L., Knappe, D. R. U., Duboscq, S., Gilmore, T. E., & Hopkins, Z. R. (2021). Per-and polyfluoroalkyl substance (PFAS) transport from groundwater to streams near a PFAS manufacturing facility in North Carolina, USA. *Environmental Science & Technology, 55*(9), 5848–5856.

Plumlee, M. H., McNeill, K., & Reinhard, M. (2009). Indirect photolysis of perfluorochemicals: Hydroxyl radical-initiated oxidation of N-ethyl perfluorooctane sulfonamido acetate (N-EtFOSAA) and other perfluoroalkanesulfonamides. *Environmental Science & Technology, 43*(10), 3662–3668.

Post, G. B., Cohn, P.D., Cooper, K.R. (2012). Perfluorinated chemicals (PFCs) – Emerging drinking water contaminants: A critical review of recent literature. Environmental Research, 116, 93-117

Post, G. B., Cohn, P. D., & Cooper, K. R. (2012). Perfluorooctanoic acid (PFOA), an emerging drinking water contaminant: A critical review of recent literature. *Environmental Research, 116*, 93–117. https://doi.org/https://doi.org/10.1016/j.envres.2012.03.007

Prevedouros, K., Cousins, I. T., Buck, R. C., & Korzeniowski, S. H. (2006). Sources, fate and transport of perfluorocarboxylates. *Environmental Science & Technology, 40*(1), 32–44.

Rankin, K., Lee, H., Tseng, P. J., & Mabury, S. A. (2014). Investigating the biodegradability of a fluorotelomer-based acrylate polymer in a soil–plant microcosm by indirect and direct analysis. *Environmental Science & Technology, 48*(21), 12783–12790.

Rankin, K., Mabury, S. A., Jenkins, T. M., & Washington, J. W. (2016). A North American and global survey of perfluoroalkyl substances in surface soils: Distribution patterns and mode of occurrence. *Chemosphere, 161*, 333–341.

Redmon, J. H., DeLuca, N. M., Thorp, E., Liyanapatirana, C., Allen, L., Andrew J., & Kondash, A. J. (2025). Hold my beer: The linkage between municipal water and brewing location on PFAS in popular beverages. *Environmental Science & Technology, 59*(17), 8368–8379. https://doi.org/10.1021/acs.est.4c11265

Rhoads, K. R., Janssen, E. M.-L., Luthy, R. G., & Criddle, C. S. (2008). Aerobic biotransformation and fate of N-ethyl perfluorooctane sulfonamidoethanol (N-EtFOSE) in activated sludge. *Environmental Science & Technology*, 42(8), 2873–2878.

Riedel, T. P., Wallace, M. A. G., Shields, E. P., Ryan, J. V., Lee, C. W., & Linak, W. P. (2021). Low temperature thermal treatment of gas-phase fluorotelomer alcohols by calcium oxide. *Chemosphere*, 272, 129859. https://doi.org/https://doi.org/10.1016/j.chemosphere.2021.129859

Roostaei, J., Colley, S., Mulhern, R., May, A. A., & Gibson, J. M. (2021). Predicting the risk of GenX contamination in private well water using a machine-learned Bayesian network model. *Journal of Hazardous Materials*, 411, 125075. https://doi.org/https://doi.org/10.1016/j.jhazmat.2021.125075

Roth, J., Abusallout, I., Hill, T., Holton, C., Thapa, U., & Hanigan, D. (2020a). Release of volatile per-and polyfluoroalkyl substances from aqueous film-forming foam. *Environmental Science & Technology Letters*, 7(3), 164–170.

Roth, J., Abusallout, I., Hill, T., Holton, C., Thapa, U., & Hanigan, D. (2020b). Response to comment on "release of volatile per-and polyfluoroalkyl substances from aqueous film-forming foam." *Environmental Science & Technology Letters*, 7(11), 869–870.

Routti, H., Aars, J., Fuglei, E., Hanssen, L., Lone, K., Polder, A., Pedersen, Å. Ø., Tartu, S., Welker, J. M., & Yoccoz, N. G. (2017). Emission changes dwarf the influence of feeding habits on temporal trends of per-and polyfluoroalkyl substances in two Arctic top predators. *Environmental Science & Technology*, 51(20), 11996–12006.

Rovero, M., Cutt, D., Griffiths, R., Filipowicz, U., Mishkin, K., White, B., Goodrow, S., & Wilkin, R. T. (2021). Limitations of current approaches for predicting groundwater vulnerability from PFAS contamination in the vadose zone. *Groundwater Monitoring & Remediation*, 41(4), 62–75.

Russell, M. H., Berti, W. R., Szostek, B., Wang, N., & Buck, R. C. (2010). Evaluation of PFO formation from the biodegradation of a fluorotelomer-based urethane polymer product in aerobic soils. *Polymer Degradation and Stability*, 95(1), 79–85.

Sáez, M., De Voogt, P., & Parsons, J. R. (2008). Persistence of perfluoroalkylated substances in closed bottle tests with municipal sewage sludge. *Environmental Science and Pollution Research*, 15, 472–477.

Schaefer, C. E., Culina, V., Nguyen, D., & Field, J. (2019). Uptake of poly-and perfluoroalkyl substances at the air–water interface. *Environmental Science & Technology*, 53(21), 12442–12448.

Schaefer, C. E., Hooper, J., Modiri-Gharehveran, M., Drennan, D. M., Beecher, N., & Lee, L. (2022). Release of poly-and perfluoroalkyl substances from finished biosolids in soil mesocosms. *Water Research*, 217, 118405.

Schaefer, C. E., Nguyen, D., Christie, E., Shea, S., Higgins, C. P., & Field, J. A. (2021). Desorption of poly-and perfluoroalkyl substances from soil historically impacted with aqueous film-forming foam. *Journal of Environmental Engineering*, 147(2), 6020006.

Schaider, L. A., Balan, S. A., Blum, A., Andrews, D. Q., Strynar, M. J., Dickinson, M. E., Lunderberg, D. M., Lang, J. R., & Peaslee, G. F. (2017). Fluorinated compounds in US fast food packaging. *Environmental Science & Technology Letters*, 4(3), 105–111.

Schenker, U., Scheringer, M., Macleod, M., Martin, J. W., Cousins, I. T., & Hungerbühler, K. (2008). Contribution of volatile precursor substances to the flux of perfluorooctanoate to the Arctic. *Environmental Science & Technology*, 42(10), 3710–3716.

Schroeder, T., Bond, D., & Foley, J. (2021). PFAS soil and groundwater contamination: Via industrial airborne emission and land deposition in SW Vermont and Eastern New York State, USA. *Environmental Science: Processes and Impacts*, 23(2), 291–301. https://doi.org/10.1039/d0em00427h

Schwichtenberg, T., Bogdan, D., Carignan, C. C., Reardon, P., Rewerts, J., Wanzek, T., & Field, J. A. (2020). PFAS and dissolved organic carbon enrichment in surface water foams on a northern US freshwater lake. *Environmental Science & Technology, 54*(22), 14455–14464.

Sehmel, G. A. (1984). *Deposition and resuspension. Atmospheric science and power production.* https://doi.org/Report DOE/TIC-2760

Sepulvado, J. G., Blaine, A. C., Hundal, L. S., & Higgins, C. P. (2011). Occurrence and fate of perfluorochemicals in soil following the land application of municipal biosolids. *Environmental Science & Technology, 45*(19), 8106–8112.

Services, W.-E. F. (2022). *Final data gap investigation report, data gap investigations and streamlined EE/CAs to support non-time critical removal actions, sites AFFF area 1 and AFFF area 21, Wright-Patterson Air Force Base, Dayton, Montgomery County, Ohio.*

SES, A. (2021). *Final remedial design/remedial action work plan, FT002 at Clarks Marsh Interim Remedial Action, Former Wurtsmith Air Force Base, Oscada, Michigan.*

Shan, G., Wei, M., Zhu, L., Liu, Z., & Zhang, Y. (2014). Concentration profiles and spatial distribution of perfluoroalkyl substances in an industrial center with condensed fluorochemical facilities. *Science of the Total Environment, 490*, 351–359.

Sharifan, H., Bagheri, M., Wang, D., Burken, J. G., Higgins, C. P., Liang, Y., Liu, J., Schaefer, C. E., & Blotevogel, J. (2021). Fate and transport of per-and polyfluoroalkyl substances (PFASs) in the vadose zone. *Science of the Total Environment, 771*, 145427.

Shi, Y., Vestergren, R., Zhou, Z., Song, X., Xu, L., Liang, Y., & Cai, Y. (2015). Tissue distribution and whole body burden of the chlorinated polyfluoroalkyl ether sulfonic acid F-53B in crucian carp (Carassius carassius): Evidence for a highly bioaccumulative contaminant of emerging concern. *Environmental Science & Technology, 49*(24), 14156–14165.

Shin, H.-M., Ryan, P. B., Vieira, V. M., & Bartell, S. M. (2012). Modeling the air–soil transport pathway of perfluorooctanoic acid in the mid-Ohio Valley using linked air dispersion and vadose zone models. *Atmospheric Environment, 51*, 67–74.

Silva, J. A. K., Martin, W. A., & McCray, J. E. (2021). Air-water interfacial adsorption coefficients for PFAS when present as a multi-component mixture. *Journal of Contaminant Hydrology, 236*, 103731.

Silva, J. A. K., Martin, W. A., Johnson, J. L., & McCray, J. E. (2019). Evaluating air-water and NAPL-water interfacial adsorption and retention of perfluorocarboxylic acids within the Vadose zone. *Journal of Contaminant Hydrology, 223*, 103472.

Sima, M. W., & Jaffé, P. R. (2021). A critical review of modeling poly-and perfluoroalkyl substances (PFAS) in the soil-water environment. *Science of the Total Environment, 757*, 143793.

Skutlarek, D., Exner, M., & Farber, H. (2006). Perfluorinated surfactants in surface and drinking waters. *Environmental Science and Pollution Research International, 13*(5), 299.

Slinn, W. G. N. (1984). *Precipitation scavenging. atmospheric science and power production.* https://doi.org/Report DOE/TIC-27601

Stahl, L. L., Snyder, B. D., Olsen, A. R., Kincaid, T. M., Wathen, J. B., & McCarty, H. B. (2014). Perfluorinated compounds in fish from US urban rivers and the Great Lakes. *Science of the Total Environment, 499*, 185–195.

Stahl, T, Heyn, J., Thiele, H., Hüther, J., Failing, K., Georgii, S., & Brunn, H. (2009). Carryover of perfluorooctanoic acid (PFOA) and perfluorooctane sulfonate (PFOS) from soil to plants. *Archives of Environmental Contamination and Toxicology, 57*, 289–298.

Stahl, Thorsten, Riebe, R. A., Falk, S., Failing, K., & Brunn, H. (2013). Long-term lysimeter experiment to investigate the leaching of perfluoroalkyl substances (PFASs) and the carry-over from soil to plants: Results of a pilot study. *Journal of Agricultural and Food Chemistry, 61*(8), 1784–1793.

Steffens, S. D., Cook, E. K., Sedlak, D. L., & Alvarez-Cohen, L. (2021). Under-reporting potential of perfluorooctanesulfonic acid (PFOS) under high-ionic strength conditions. *Environmental Science & Technology Letters*, *8*(12), 1032–1037.

Stoiber, T., Evans, S., & Naidenko, O. V. (2020). Disposal of products and materials containing per-and polyfluoroalkyl substances (PFAS): A cyclical problem. *Chemosphere*, *260*, 127659.

Szabo, D., Nuske, M. R., Lavers, J. L., Shimeta, J., Green, M. P., Mulder, R. A., & Clarke, B. O. (2022). A baseline study of per-and polyfluoroalkyl substances (PFASs) in waterfowl from a remote Australian environment. *Science of the Total Environment*, *812*, 152528.

Taniyasu, S., Yamashita, N., Moon, H.-B., Kwok, K. Y., Lam, P. K. S., Horii, Y., Petrick, G., & Kannan, K. (2013). Does wet precipitation represent local and regional atmospheric transportation by perfluorinated alkyl substances? *Environment International*, *55*, 25–32. https://doi.org/https://doi.org/10.1016/j.envint.2013.02.005

Tech, T. (2022). Final annual technical memorandum for per and polyfluoroalkyl substances, July and October 2019, January and May 2020 Surface Water Sampling and July 2019 Sediment Sampling.

Thackray, C. P., & Selin, N. E. (2017). Uncertainty and variability in atmospheric formation of PFCAs from fluorotelomer precursors. *Atmospheric Chemistry and Physics*, *17*(7), 4585–4597.

Thackray, C. P., Selin, N. E., & Young, C. J. (2020). A global atmospheric chemistry model for the fate and transport of PFCAs and their precursors. *Environmental Science: Processes & Impacts*, *22*(2), 285–293.

Tian, Y., Yao, Y., Chang, S., Zhao, Z., Zhao, Y., Yuan, X., Wu, F., & Sun, H. (2018). Occurrence and phase distribution of neutral and ionizable per-and polyfluoroalkyl substances (PFASs) in the atmosphere and plant leaves around landfills: A case study in Tianjin, China. *Environmental Science & Technology*, *52*(3), 1301–1310.

Titaley, I. A., De la Cruz, F. B., & Field, J. A. (2020). Comment on "release of volatile per-and polyfluoroalkyl substances from aqueous film-forming foam." *Environmental Science & Technology Letters*, *7*(11), 866–868.

Tokranov, A. K., LeBlanc, D. R., Pickard, H. M., Ruyle, B. J., Barber, L. B., Hull, R. B., Sunderland, E. M., & Vecitis, C. D. (2021). Surface-water/groundwater boundaries affect seasonal PFAS concentrations and PFAA precursor transformations. *Environmental Science: Processes & Impacts*, *23*(12), 1893–1905.

Tomy, G. T., Budakowski, W., Halldorson, T., Helm, P. A., Stern, G. A., Friesen, K., Pepper, K., Tittlemier, S. A., & Fisk, A. T. (2004). Fluorinated organic compounds in an eastern Arctic marine food web. *Environmental Science & Technology*, *38*(24), 6475–6481.

Tomy, G. T., Pleskach, K., Ferguson, S. H., Hare, J., Stern, G., MacInnis, G., Marvin, C. H., & Loseto, L. (2009). Trophodynamics of some PFCs and BFRs in a western Canadian Arctic marine food web. *Environmental Science & Technology*, *43*(11), 4076–4081.

Tseng, N., Wang, N., Szostek, B., & Mahendra, S. (2014). Biotransformation of 6: 2 fluorotelomer alcohol (6: 2 FTOH) by a wood-rotting fungus. *Environmental Science & Technology*, *48*(7), 4012–4020.

USEPA. (2016b). *Ecological effects tests guidelines OCSPP 850.1730: Fish Bioconcentration Factor (BCF)*. https://doi.org/EPA 712-C-16-003

USEPA. (2000). Perfluorooctyl sulfonates; proposed significant new use rule. In Fed Reg, 65. Available at: https://www.federalregister.gov/documents/2000/10/18/00-26751/perfluorooctyl-sulfonates-proposed-significant-new-use-rule

USEPA. (2016a). *AERMOD implementation guide*. https://doi.org/EPA-454/B-16-013

USEPA. (2017a). *Assessing and managing chemicals under TSCA*. Last Modified December 18, 2019. accessed 2020.

USEPA. (2017d). *Technical fact sheet–perfluorooctane sulfonate (PFOS) and perfluorooctanoic acid (PFOA)*. USEPA.

Vedagiri, U. K., Anderson, R. H., Loso, H. M., & Schwach, C. M. (2018). Ambient levels of PFOS and PFOA in multiple environmental media. *Remediation Journal*, 28(2), 9–51.

Venkatesan, A. K., & Halden, R. U. (2013). National inventory of perfluoroalkyl substances in archived US biosolids from the 2001 EPA National Sewage Sludge Survey. *Journal of Hazardous Materials*, 252, 413–418.

Wagner, A., Raue, B., Brauch, H.-J., Worch, E., & Lange, F. T. (2013). Determination of adsorbable organic fluorine from aqueous environmental samples by adsorption to polystyrene-divinylbenzene based activated carbon and combustion ion chromatography. *Journal of Chromatography A*, 1295, 82–89.

Wang, N., Liu, J., Buck, R. C., Korzeniowski, S. H., Wolstenholme, B. W., Folsom, P. W., & Sulecki, L. M. (2011). 6: 2 Fluorotelomer sulfonate aerobic biotransformation in activated sludge of waste water treatment plants. *Chemosphere*, 82(6), 853–858.

Wang, N., Szostek, B., Buck, R. C., Folsom, P. W., Sulecki, L. M., & Gannon, J. T. (2009). 8-2 Fluorotelomer alcohol aerobic soil biodegradation: Pathways, metabolites, and metabolite yields. *Chemosphere*, 75(8), 1089–1096.

Wang, N., Szostek, B., Buck, R. C., Folsom, P. W., Sulecki, L. M., Capka, V., Berti, W. R., & Gannon, J. T. (2005b). Fluorotelomer alcohol biodegradation direct evidence that perfluorinated carbon chains breakdown. *Environmental Science & Technology*, 39(19), 7516–7528.

Wang, N., Szostek, B., Folsom, P. W., Sulecki, L. M., Capka, V., Buck, R. C., Berti, W. R., & Gannon, J. T. (2005a). Aerobic biotransformation of 14C-labeled 8-2 telomer B alcohol by activated sludge from a domestic sewage treatment plant. *Environmental Science & Technology*, 39(2), 531–538.

Wang, W., Rhodes, G., Ge, J., Yu, X., & Li, H. (2020). Uptake and accumulation of per-and polyfluoroalkyl substances in plants. *Chemosphere*, 261, 127584.

Wang, X., Halsall, C., Codling, G., Xie, Z., Xu, B., Zhao, Z., Xue, Y., Ebinghaus, R., & Jones, K. C. (2014). Accumulation of perfluoroalkyl compounds in tibetan mountain snow: Temporal patterns from 1980 to 2010. *Environmental Science & Technology*, 48(1), 173–181.

Wang, Z., DeWitt, J. C., Higgins, C. P., & Cousins, I. T. (2017a). *A never-ending story of per-and polyfluoroalkyl substances (PFASs)?* ACS Publications.

Wang, Z., Xie, Z., Mi, W., Möller, A., Wolschke, H., & Ebinghaus, R. (2015). Neutral poly/per-fluoroalkyl substances in air from the Atlantic to the Southern Ocean and in Antarctic Snow. *Environmental Science & Technology*, 49(13), 7770–7775. https://doi .org/10.1021/acs.est.5b00920

Wania, F. (2007). A global mass balance analysis of the source of perfluorocarboxylic acids in the Arctic Ocean. *Environmental Science & Technology*, 41(13), 4529–4535.

Washington, J. W., Ellington, J. J., Jenkins, T. M., Evans, J. J., Yoo, H., & Hafner, S. C. (2009). Degradability of an acrylate-linked, fluorotelomer polymer in soil. *Environmental Science & Technology*, 43(17), 6617–6623.

Washington, J. W., Jenkins, T. M., Rankin, K., & Naile, J. E. (2015). Decades-scale degradation of commercial, side-chain, fluorotelomer-based polymers in soils and water. *Environmental Science & Technology*, 49(2), 915–923.

Washington, J. W., Naile, J. E., Jenkins, T. M., & Lynch, D. G. (2014). Characterizing fluorotelomer and polyfluoroalkyl substances in new and aged fluorotelomer-based polymers for degradation studies with GC/MS and LC/MS/MS. *Environmental Science & Technology*, 48(10), 5762–5769.

Washington, J. W., Rankin, K., Libelo, E. L., Lynch, D. G., & Cyterski, M. (2019). Determining global background soil PFAS loads and the fluorotelomer-based polymer degradation rates that can account for these loads. *Science of the Total Environment*, 651, 2444–2449.

Washington, J. W., Yoo, H., Ellington, J. J., Jenkins, T. M., & Libelo, E. L. (2010). Concentrations, distribution, and persistence of perfluoroalkylates in sludge-applied soils near Decatur, Alabama, USA. *Environmental Science & Technology*, *44*(22), 8390–8396.

Weber, A. K., Barber, L. B., LeBlanc, D. R., Sunderland, E. M., & Vecitis, C. D. (2017). Geochemical and hydrologic factors controlling subsurface transport of poly-and perfluoroalkyl substances, Cape Cod, Massachusetts. *Environmental Science & Technology*, *51*(8), 4269–4279.

Wei, C., Song, X., Wang, Q., & Hu, Z. (2017). Sorption kinetics, isotherms and mechanisms of PFOS on soils with different physicochemical properties. *Ecotoxicology and Environmental Safety*, *142*, 40–50.

Weiner, B., Yeung, L. W. Y., Marchington, E. B., D'Agostino, L. A., & Mabury, S. A. (2013). Organic fluorine content in aqueous film forming foams (AFFFs) and biodegradation of the foam component 6: 2 fluorotelomermercaptoalkylamido sulfonate (6: 2 FTSAS). *Environmental Chemistry*, *10*(6), 486–493.

Wen, B., Li, L., Liu, Y., Zhang, H., Hu, X., Shan, X., & Zhang, S. (2013). Mechanistic studies of perfluorooctane sulfonate, perfluorooctanoic acid uptake by maize (Zea mays L. cv. TY2). *Plant and Soil*, *370*, 345–354.

White, N. D., Balthis, L., Kannan, K., De Silva, A. O., Wu, Q., French, K. M., Daugomah, J., Spencer, C., & Fair, P. A. (2015). Elevated levels of perfluoroalkyl substances in estuarine sediments of Charleston, SC. *Science of the Total Environment*, *521*, 79–89.

Wood. (2020). Final expanded site inspection report for per- and polyfluoroalkyl substances, 105th Airlift Wing, New York Air National Guard, Stewart Air National Guard Base, Newburgh, NY. U.S. Air National Guard.

Xiao, F., Jin, B., Golovko, S. A., Golovko, M. Y., & Xing, B. (2019). Sorption and desorption mechanisms of cationic and zwitterionic per-and polyfluoroalkyl substances in natural soils: Thermodynamics and hysteresis. *Environmental Science & Technology*, *53*(20), 11818–11827.

Xiao, F., Simcik, M. F., Halbach, T. R., & Gulliver, J. S. (2015). Perfluorooctane sulfonate (PFOS) and perfluorooctanoate (PFOA) in soils and groundwater of a US metropolitan area: Migration and implications for human exposure. *Water Research*, *72*, 64–74.

Xiao, X., Ulrich, B. A., Chen, B., & Higgins, C. P. (2017a). Sorption of poly-and perfluoroalkyl substances (PFASs) relevant to aqueous film-forming foam (AFFF)-impacted groundwater by biochars and activated carbon. *Environmental Science & Technology*, *51*(11), 6342–6351.

Yan, P.-F., Dong, S., Manz, K. E., Liu, C., Woodcock, M. J., Mezzari, M. P., Abriola, L. M., Pennell, K. D., & Cápiro, N. L. (2022). Biotransformation of 8: 2 fluorotelomer alcohol in soil from aqueous film-forming foams (AFFFs)-impacted sites under nitrate-, sulfate-, and iron-reducing conditions. *Environmental Science & Technology*, *56*(19), 13728–13739.

Yarwood, G., Kemball-Cook, S., Keinath, M., Waterland, R. L., Korzeniowski, S. H., Buck, R. C., Russell, M. H., & Washburn, S. T. (2007). High-resolution atmospheric modeling of fluorotelomer alcohols and perfluorocarboxylic acids in the North American troposphere. *Environmental Science & Technology*, *41*(16), 5756–5762.

Yeung, L. W. Y., Dassuncao, C., Mabury, S., Sunderland, E. M., Zhang, X., & Lohmann, R. (2017). Vertical profiles, sources, and transport of PFASs in the Arctic Ocean. *Environmental Science & Technology*, *51*(12), 6735–6744.

Yi, S., Harding-Marjanovic, K. C., Houtz, E. F., Antell, E., Olivares, C., Nichiporuk, R. V, Iavarone, A. T., Zhuang, W.-Q., Field, J. A., & Sedlak, D. L. (2022). Biotransformation of 6: 2 fluorotelomer thioether amido sulfonate in aqueous film-forming foams under nitrate-reducing conditions. *Environmental Science & Technology*, *56*(15), 10646–10655.

Yi, S., Harding-Marjanovic, K. C., Houtz, E. F., Gao, Y., Lawrence, J. E., Nichiporuk, R. V, Iavarone, A. T., Zhuang, W.-Q., Hansen, M., & Field, J. A. (2018). Biotransformation of AFFF component 6: 2 fluorotelomer thioether amido sulfonate generates 6: 2 fluorotelomer thioether carboxylate under sulfate-reducing conditions. *Environmental Science & Technology Letters*, 5(5), 283–288.

Young, C. J., & Mabury, S. A. (2010). Atmospheric perfluorinated acid precursors: Chemistry, occurrence, and impacts. In P. De Voogt (Ed.), Reviews of environmental contamination and toxicology: *Perfluorinated alkylated substances* (Vol. 208, pp. 1–109). Springer.

Yu, Q., Zhang, R., Deng, S., Huang, J., & Yu, G. (2009). Sorption of perfluorooctane sulfonate and perfluorooctanoate on activated carbons and resin: Kinetic and isotherm study. *Water Research*, 43(4), 1150–1158.

Zeng, J., & Guo, B. (2021). Multidimensional simulation of PFAS transport and leaching in the vadose zone: Impact of surfactant-induced flow and subsurface heterogeneities. *Advances in Water Resources*, 155, 104015.

Zhang, H., Liu, W., He, X., Wang, Y., & Zhang, Q. (2015b). Uptake of perfluoroalkyl acids in the leaves of coniferous and deciduous broad-leaved trees. *Environmental Toxicology and Chemistry*, 34(7), 1499–1504.

Zhang, L., Lee, L. S., Niu, J., & Liu, J. (2017). Kinetic analysis of aerobic biotransformation pathways of a perfluorooctane sulfonate (PFOS) precursor in distinctly different soils. *Environmental Pollution*, 229, 159–167.

Zhang, L., Sun, H., Wang, Q., Chen, H., Yao, Y., Zhao, Z., & Alder, A. C. (2019). Uptake mechanisms of perfluoroalkyl acids with different carbon chain lengths (C2-C8) by wheat (Triticum acstivnm L.). *Science of the Total Environment*, 654, 19–27.

Zhang, R., Yan, W., & Jing, C. (2015a). Experimental and molecular dynamic simulation study of perfluorooctane sulfonate adsorption on soil and sediment components. *Journal of Environmental Sciences*, 29, 131–138.

Zhang, S., Szostek, B., McCausland, P. K., Wolstenholme, B. W., Lu, X., Wang, N., & Buck, R. C. (2013). 6: 2 and 8: 2 fluorotelomer alcohol anaerobic biotransformation in digester sludge from a WWTP under methanogenic conditions. *Environmental Science & Technology*, 47(9), 4227–4235.

Zhao, L., Zhou, M., Zhang, T., & Sun, H. (2013). Polyfluorinated and perfluorinated chemicals in precipitation and runoff from cities across Eastern and Central China. *Archives of Environmental Contamination and Toxicology*, 64(2), 198–207. https://doi.org/10.1007/s00244-012-9832-x

Zhi, Y., & Liu, J. (2018). Sorption and desorption of anionic, cationic and zwitterionic polyfluoroalkyl substances by soil organic matter and pyrogenic carbonaceous materials. *Chemical Engineering Journal*, 346, 682–691.

Zhong, W., Zhang, L., Cui, Y., Chen, M., & Zhu, L. (2019). Probing mechanisms for bioaccumulation of perfluoroalkyl acids in carp (Cyprinus carpio): Impacts of protein binding affinities and elimination pathways. *Science of the Total Environment*, 647, 992–999.

Zhou, J., Baumann, K., Mead, R. N., Skrabal, S. A., Kieber, R. J., Avery, G. B., Shimizu, M., Dewitt, J. C., Sun, M., Vance, S. A., Bodnar, W., Zhang, Z., Collins, L. B., Surratt, J. D., & Turpin, B. J. (2021). PFOS dominates PFAS composition in ambient fine particulate matter (PM2.5) collected across North Carolina nearly 20 years after the end of its US production. *Environmental Science: Processes and Impacts*, 23(4), 580–587. https://doi.org/10.1039/d0em00497a

4 Regulations and Guidelines for Perfluoroalkyl and Polyfluoroalkyl Substances

4.1 INTRODUCTION

Given its anti-grease, water-resistant, and stain-repelling qualities, per- and poly-fluoroalkyl substances (PFAS) are made by human chemical compounds (Glüge et al., 2020; Dawson et al., 2023; Redmon et al., 2025). For many years, PFAS have been extensively utilized in industrial manufacturing (e.g., as a processing aid in the manufacturing of plastic), consumer goods (CGs) (e.g., carpets, furniture, clothing, kitchenware, and food packaging), and firefighting foam (FFF) (Glüge et al., 2020). Because of their widespread utilization and ability to withstand degradation, they are frequently referred to as "forever chemicals" and are found throughout the environment. According to recent research, exposure to PFAS can have detrimental consequences on the body's development, reproduction, cardiovascular system, liver, kidneys, immune system, and cancer (Redmon et al., 2025). PFAS can also linger in the human body for several decades (Dawson et al., 2023).

PFAS are being handled by states and countries worldwide in several ways, including restricting or banning their applications in common source substances, reporting requirements, enforcing testing, and funding of repairs as well as monitoring. A summary of each continent's PFAS laws is provided in this paragraph. Figure 4.1 presents a graphical depiction of the legislative chronology.

Particularly outside of Europe, the regulations of PFAS now in place to control the levels of PFAS in food and drinking water are deficient. Tables 4.1, 4.2, and 4.3 provide an overview of the present regulations of PFAS (EEA, 2023a; Health Canada, 2023a, 2023b; NHMRC, 2022; US.EPA, 2024a; Wang et al., 2023).

DOI: 10.1201/9781003625537-4

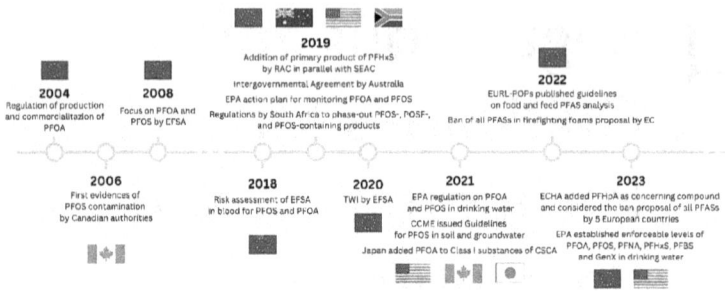

FIGURE 4.1   Timeline of PFAS legislation.

TABLE 4.1
Allowed values of PFAS (µg/L) in drinking water

Compound	US	Europe	Canada	Japan	China	Australia
PFOS	0.0040	N	N	0.050	0.040	0.070*
PFOA	0.0040	N	N	0.050	0.080	0.56
PFHxS	0.0010	N	N	N	N	0.070*
PFNA	0.0010	N	N	N	N	NOL
Total PFAS	NOL	0.50	0.030	N	N	NOL

(N = not legislated; * = Total of PFOS and PFHxS).

TABLE 4.2
Levels of PFAS (µg/kg) allowed in eggs

Compound	America	Europe	Asia	Oceania	Africa
PFOA	N	1.0	N	N	N
PFNA	N	0.70	N	N	N
PFOS	N	0.30	N	N	N
PFHxS	N	0.30	N	N	N
Total PFAS	N	N	N	N	N

(N = not legislated)

4.2   EUROPEAN UNION PFAS MONITORING

In 2004, significant European agencies started the process of monitoring PFAS pollution. The European Union released a law this year that governed the manufacture and sale of PFOS and that was in effect until 2019 (EEA, 2023b). The "Regulation European Commission (EU) 2019/1021 of the European Parliament and of the Council of June 20, 2019 on persistent organic pollutants" took the place of that

TABLE 4.3

Permitted quantities of PFAS (µg/kg) in fish flesh

Compound	America	Europe	Asia	Oceania	Africa
PFHxS	N	0.20	N	N	N
PFOA	N	2.0	N	N	N
PFNA	N	2.5	N	N	N
PFOS	N	1.0	N	N	N
Total PFAS	N	N	N	N	N

(N = not legislated; * = Total of PFOS and PFHxS).

legislative instrument (EEA, 2023c), where it is forbidden, with some exceptions, to manufacture, utilize, and put on the market any materials included in Annex I. This refers to pollutants such as PFOS as well as its derivatives $C_8F_{17}SO_2X$ (X = OH, metal salt [O-M+]), amide, halide, as well as other derivatives together with polymers, in mixes or products. In 2008, PFOA, PFOS, and their salts (byproducts of the degradation of PFAS in polymers) were the subject of a scientific opinion published by EFSA's panel on pollutants in the food chain. The view resolved that there is still uncertainty regarding the proportional involvement of these precursors to the ecological/environmental pollution caused by PFOS and PFOA (Benford et al., 2008). The risk assessment on human health associated with PFOS and PFOA in food was later released by EFSA in 2018, before the announcement by one of the world's foremost independent research and technology organisations, with expertise in materials joining and engineering processes (known as TWI Ltd) . The evaluation said that "PFOS and PFOA were detected in blood samples of almost all individuals assessed, demonstrating ubiquitous exposure". This has already shown that new analytical techniques and legislative advances are temporarily required (Knutsen et al., 2018). The most recent analytical methods were enumerated in the same view.

The addition of PFHxS (and its salts) and compounds linked to it to "Annexes A, B, and/or C of the Stockholm Convention on POPs" was suggested by the Norwegian Environment Agency in 2019. Around 150 chemicals that might degrade into PFHxS are included in the dossier submitter's recommended scope that the Committee for Risk Assessment (RAC) contributed concurrently. Meanwhile, the Socio-Economic Analysis Committee (SEAC) concluded that an additional comprehensive approach is needed in addressing the risks related to products as well as the mixtures encompassing PFHxS, its salts, and associated chemicals, instead of relying exclusively on national regulations as contained in the European Chemicals Agency (ECHA) document (ECHA, 2020). The EU suggested a prohibition on all PFAS in FFFs across the European Union on February 23, 2022, citing rising contamination levels and technological advancements that allow them to be detected at lower levels. Such a limitation is intended to combat hazards to human health and soil and groundwater pollution. To lessen the dangers associated with PFAS in FFFs, the ECHA evaluated the advantages and disadvantages of five potential approaches (ECHA, 2022b).

The Annex XV Restriction Report for PFAS in FFFs published in March 2022 proposes to ban the marketing, utilization, as well as formulation or preparation of any PFAS in post-use FFFs or industry-specific transition periods (ECHA, 2022a). Also, Perfluoropentanoic acid (PFHpA) was added to the candidate list of chemicals of extremely high concern by the ECHA on January 17, 2023, after the suggestions for the prohibition of PFOA, PFOS, and PFHxS. This was due to the compound's harmful qualities. Despite not being registered with REACH, this compound's listing might be viewed as a preventative measure to avoid using it as a potential future PFAS replacement (ECHA, 2023a).

When Germany, Denmark, Norway, the Netherlands, and Sweden proposed to ECHA that PFAS be banned under REACH, it was the most momentous move in the European Union's fight against pollution by per- and polyfluoroalkyl substances. The European Economic Area (EEA) and EU need to address the hazards that the five agencies discovered are associated with the improper control, production, marketing, and use of the pollutants. On February 7, 2023, ECHA released the comprehensive plan, among the most extensive in EU history. During their meetings in March 2023, RAC and SEAC of ECHA deliberated whether or not the proposal satisfies the legislative standards of REACH. In that case, the committees started the evaluation of the proposal scientifically. About a half-year dialogue was scheduled to begin on March 22, 2023 (ECHA, 2023b).

4.3   THE AMERICAS

4.3.1   United States (US)

In the state legislatures, the US and the federal government are taking actions that will protect and safeguard public health as well as the environment; both terrestrial and aquatic environments (NCSL, 2023). Following a non-enforceable health advisory on PFOS and PFOA in 2016, the US EPA released a "formal PFAS Action Plan in 2019" and revised it in February 2020. The strategy tackles issues such as establishing maximum pollution limits for states and municipal water utilities, identifying PFOA and PFOS as hazardous compounds under CERLCA, and creating more precise and cutting-edge technology to identify soil, groundwater, and drinking water toxins. EPA released the Agency's PFAS Strategic Roadmap in October 2021 (US EPA, 2024b). This offers a more robust and protective approach and constitutes a significant and crucial step to protect communities from PFAS pollution.

The EPA began work on a national drinking water rule in March 2023 to establish maximum contamination levels (MCLs) for six perfluoroalkyl substances (PFAS): GenX compounds, PFOS, PFOA, PFNA, PFHxS, and perfluorobutane sulfonic acid (PFBS). After the final regulations go into effect, states must set their criteria at least to those levels. Legislation was concurrently enacted by state lawmakers that limited perfluoroalkyl substances (PFAS) use in FFF, controlled their presence in CGs, food packaging, and drinking water, and provided funding for cleaning and repair. More than 250 PFAS-related measures were passed by lawmakers between 2018 and 2021, regulating the substances' usage in a variety of products and establishing maximum contamination levels. In 2022, lawmakers debated about 200 proposals, including language relating to per- and polyfluoroalkyl substances (PFAS). Approximately 50 acts about PFAS in the packaging of food, personal protective equipment (PPE) for firefighters, as well as for environmental cleanup were enacted in no less than 18 states. Legislators also raised the problem of PFAS in CGs, including clothing, toys, kitchenware, furniture, cosmetics, and ski wax, in their bills. Following Maine's 2021 prohibition on the sale of goods with PFAS added on purpose, some states acquired the initiative for the regulation of PFAS.

4.3.2 CANADA

Canadian authorities began addressing PFAS in the environment in 2006 (Hains & Richards, 2024). The federal government came to the conclusion that year, that PFOS, its salts, as well as other composites can possibly enter the environment and negatively impact the environment or its biological diversity, even though the human health assessment found that the concentrations of PFOS exposure were below levels that could have an impact on human health. The chemicals listed in Canada's 2006 Prohibition of Certain Toxic Chemicals Regulations were added by the government in 2012. In 2016, Health Canada (HC) expanded drinking water testing to include seven more PFAS in addition to PFOA and PFOS. A similar soil screening require-ment was issued by HC a year later. Between 2018 and 2021, HC added two more PFAS to soil and drinking water monitoring programs and provided domestic water quality regulations for PFOS and PFOA. PFOS guidelines for groundwater and soil were published by the Canadian Council of Ministers of the Environment (CCME) in 2021. Finally, in the spring of 2021, Canada announced it would begin categoriz-ing PFAS compounds instead of individual ones.

4.3.3 BRAZIL

As part of the NIP-Brazil-2015, Brazil developed an action plan for PFOS for other South American countries. Nevertheless, some suggested actions have not yet begun or are complete (Ministry of the Environment, 2015; Torres et al., 2022). The most noteworthy accomplishment is creating a unique tracking code for PFOS and related compound foreign trade data. Its importance is highlighted given the challenge of tracking PFOA international trade statistics. Using sulfluramid in Brazilian soils is causing PFOS environmental contamination, which the Brazilian Agricultural Research Corporation is now investigating (EMBRAPA). Nonetheless, fundamental obstacles remain to be addressed, such as increasing stakeholder involvement and improving and expanding the information gathered.

Both PFOS and PFOA should have a law drafted that would provide guidelines and restrictions for the environmental accrediting, authorizing, or licensing of spe-cific PFAS usage and applications while staying within the bounds of the Stockholm Convention's permitted uses and exclusions. Evaluation of the Brazilian National Council for the Environment (CONAMA) resolutions concerning water and soil quality must include the PFAS in question in environmental quality indicators.

4.4 ASIA

Similarly, Asian countries are trying to tighten their rules (Enviliance, 2024).

4.4.1 JAPAN

In Japan, PFOA and its associated compounds were removed from the Class I Specified Chemical compounds list under the CSCA in April 2021.

4.4.2 CHINA

November 2020 saw the addition of 18 chemicals, including PFOA, to China's list of Priority Control Chemicals. This marks the beginning of legislative action aimed at controlling environmental danger. The China Ministry of Ecology and Environment published a draft plan on October 11, 2021, to better manage emerging contaminants, such as PFOA.

4.4.3 SINGAPORE

In Singapore, in 2019, PFOA was added to the list of substances that must be controlled.

4.4.4 INDONESIA

Since October 2019, Indonesia has limited the quantity of perfluoroalkyl asbestos (PFOA) in certain items, such as blankets, bedsheets, and textiles.

4.4.5 THAILAND

Eight compounds, including PFOA, were suggested as Class 4 restricted chemicals in Thailand in May 2021. A revised draft of the PFOA rules was released on February 9, 2022, by the Department of Industrial Works, Ministry of Industry of Thailand. On December 21, 2022, PFOA, its salts, and compounds linked to PFOA were added to the list of dangerous chemicals.

4.4.6 VIETNAM

The Stockholm Convention is mentioned explicitly in Vietnam's Law on Environmental Protection of 2020, for the first time. The draft decree, one of its subordinate laws, identifies PFOA as a material that has to be controlled. The Law on Environmental Protection later established permissible levels for POPs, including PFOA, in November 2021. The government of Vietnam published Decree No. 82/2022/ND-CP on October 18, 2022, which included "PFOA in ANNEX II of Decree 113/2017/ND-CP".

4.5   AFRICA

Even though several African nations have adopted national implementation plans (NIPs) and approved the Stockholm Convention, different steps are being taken to reduce exposure to PFAS (Groffen et al., 2021). Many NIPs have not been updated since they were created before 2009, or before PFAS were listed in the Convention (Ssebugere et al., 2020). PFAS are, therefore, frequently left out of these NIPs. On the other hand, some nations are revising their PFAS laws.

4.5.1   SOUTH AFRICA

For instance, by December 2021, items containing "PFOS, POSF, and PFOS" should be banned in South Africa according to legislation. Africa has only recently begun studying PFAS (Hains & Richards, 2024). From an African perspective, most countries struggle to integrate this new category of pollutants into the frameworks currently in place for environmental management. Creating ecological quality standards for hazardous chemicals is still necessary for chemical-specific management, even in nations like South Africa, where these substances are governed by well-organized environmental regulations (Claassen et al., 2020; Meyer & Roos, 2015).

4.6   OCEANIA

4.6.1   AUSTRALIA

An Intergovernmental Agreement was implemented in Australia in February 2019 to address PFAS pollution (Department of the Prime Minister and Cabinet, 2023). The agreement promotes standard solutions to PFAS adulteration that protect the environment and human health and assists Australian governments in addressing PFAS contamination. Measures must, however, be feasible, risk-adjustable, and conducive to economic stability.

4.7   BRIEF DISCUSSION ON THE GLOBAL REGULATIONS AND GUIDELINES FOR PFAS

Legislators worldwide are passing laws to regulate PFAS domestically and internationally in response to worries about the dangers PFAS pose to the environment and public health. Recently, officials and legislators in the US and Europe have put out several legislative proposals to minimize or do away with the utilization of PFAS compounds in CGs. Concurrently, countries in Asia, including Japan, China, and South Korea have initiated measures towards more stringent regulations on items containing PFAS to safeguard the health of their populace. Due to growing public awareness, many developing nations are starting to understand the necessity of laws to control PFAS and protect human health. The fact that governments are proactively attempting to lower the amount of PFAS in the environment, as well as food substances, is heartening. The total risk caused by these chemicals should be lessened in part by the limitations on the manufacturing, utilization, as well as disposal of PFAS and their composites.

4.8 CONCLUSION

A concerted international effort is needed to regulate PFAS, which is a complicated, evolving, emerging, and developing issue. This chapter briefly highlights the advancements in several regions, including Asia's increasing emphasis on PFAS monitoring, the United States' health advisory levels, and Europe's strict regulations. Despite these developments, significant obstacles still exist, such as inconsistent legislative restrictions, lax enforcement in some regions, and the requirement for standardized approaches to evaluate and reduce PFAS hazards. Future efforts should concentrate on developing breakthrough technology for PFAS detection and cleanup, strengthening cross-border cooperation, and creating consistent worldwide, rules and regulations as well as legislation of PFAS. By tackling these issues, policymakers and interested parties may better safeguard the environment and public health from the widespread effects of PFAS.

REFERENCES

Benford, D., De Boer, J., Carere, A., Di Domenico, A., Johansson, N., Schrenk, D., Schoeters, G., De Voogt, P., & Dellatte, E. (2008). Opinion of the scientific panel on contaminants in the food chain on perfluorooctane sulfonate (PFOS), perfluorooctanoic acid (PFOA) and their salts. *EFSA J, 653*, 1–131.
Claassen, M., Dabrowski, J. M., Nepfumbada, T., van der Laan, M., Shadung, J., & Thwala, M. (2020). Incorporating environmental fate models into risk assessment for pesticide registration in South Africa. *Water Research Commission: Pretoria, South Africa.*
Dawson, D. E., Lau, C., Pradeep, P., Sayre, R. R., Judson, R. S., Tornero-Velez, R., & Wambaugh, J. F. (2023). A machine learning model to estimate toxicokinetic half-lives of per-and polyfluoro-alkyl substances (PFAS) in multiple species. *Toxics, 11*(2), 98. https://doi.org/10.3390/toxics11020098
Department of the Prime Minister and Cabinet. (2023). *Intergovernmental agreement on a national framework for responding to PFAS contamination.* Department of the Prime Minister and Cabinet. https://federation.gov.au/about/agreements/intergovernmental -agreement-national-framework-responding-pfas-contamination
ECHA. (2020). *Perfluorohexane sulfonic acid (PFHxS) including its salts and related substances.* ECHA.
ECHA. (2022a). *Per- and polyfluoroalkyl substances (PFASs) in firefighting foams.* ECHA.
ECHA. (2022b). *Proposal to ban 'forever chemicals' in firefighting foams throughout the EU.* ECHA. https://echa.europa.eu/it/-/proposal-to-ban-forever-chemicals-in-firefight-ing-foams-throughout-the-eu
ECHA. (2023a). *ECHA adds nine hazardous chemicals to Candidate List.* ECHA. https://echa.europa.eu/it/-/echa-adds-nine-hazardous-chemicals-to-candidate-list
ECHA. (2023b). *ECHA receives PFASs restriction proposal from five national authorities.* ECHA. https://echa.europa.eu/it/-/echa-receives-pfass-restriction-proposal-from-five -national-authorities
EEA. (2023a). *Commission Regulation (EU) 2022/2388 of 7 December 2022 amending Regulation (EC) No 1881/2006 as regards maximum levels of perfluoroalkyl substances in certain foodstuffs.* https://eur-lex.europa.eu/eli/reg/2022/2388/oj#document1

EEA. (2023b). *Regulation (EC) No 850/2004 of the European Parliament and of the Council of 29 April 2004 on persistent organic pollutants and amending Directive 79/117/EEC.* https://eur-lex.europa.eu/legal-content/EN/ALL/?uri=celex:32004R0850

EEA. (2023c). *Regulation (EU) 2019/1021 of the European Parliament and of the Council of 20 June 2019 on persistent organic pollutants (recast) (Text with EEA relevance.).* https://eur-lex.europa.eu/legal-content/en/TXT/?uri=CELEX%3A32019R1021

Enviliance. (2024). *PFOA regulations in Asian countries.* Enviliance. https://enviliance.com/regions/others/asia-pfoa

Glüge, J., Scheringer, M., Cousins, I. T., DeWitt, J. C., Goldenman, G., Herzke, D., Lohmann, R., Ng, C. A., Trier, X., & Wang, Z. (2020). An overview of the uses of per-and polyfluoroalkyl substances (PFAS). *Environmental Science: Processes & Impacts, 22*(12), 2345–2373. https://doi.org/10.1039/D0EM00291G

Groffen, T., Nkuba, B., Wepener, V., & Bervoets, L. (2021). Risks posed by per- and polyfluoroalkyl substances (PFAS) on the African continent, emphasizing aquatic ecosystems. *Integrated Environmental Assessment and Management, 17*(4), 726–732.

Hains, J., & Richards, S. (2024). *PFAS in Canadian provinces: Where are the regulations?* Stantec. https://www.stantec.com/en/ideas/topic/covid-19/pfas-in-canadian-provinces-where-are-the-regulations#:~:text=federal government doing%3F-, On a federal level%2C Canada took steps to deal with, environment or its biological diversity.

Health Canada. (2023a). *Water talk: Per-and polyfluoroalkyl substances (PFAS) in drinking water.* Canada.Ca. https://www.canada.ca/en/health-canada/services/environmental-workplace-health/reports-publications/water-quality/water-talk-per-polyfluoroalkyl-substances-drinking-water.html

Health Canada. (2023b). *Water talk: Per-and polyfluoroalkyl substances (PFAS) in drinking water.* https://www.canada.ca/en/health-canada/services/environmental-workplace-health/reports-publications/water-quality/water-talk-per-polyfluoroalkyl-substances-drinking-water.html

Knutsen, H. K., Alexander, J., Barregård, L., Bignami, M., Brüschweiler, B., & Ceccatelli, S. (2018). Risk to human health related to the presence of perfluorooctane sulfonic acid and perfluorooctanoic acid in food. *EFSA J, 16*(12), e05194, PMID: 32625773.

Meyer, T., & Roos, C. (2015). Regulation and management of hazardous chemical substances in South Africa. In: Newman, M. (Ed.) *Fundamentals of Ecotoxicology: The Science of Pollution*, 4th Ed.; Newman, M., Ed, 463–469.

Ministry of the Environment. (2015). *National implementation plan Brazil: Convention Stockholm/Ministry of the Environment.*

NCSL. (2023). *Per- and polyfluoroalkyl substances (PFAS) | State Legislation and Federal Action.* NCSL. https://www.ncsl.org/environment-and-natural-resources/per-and-polyfluoroalkyl-substances

NHMRC. (2022). *Australian drinking water guidelines version 3.8.*

Redmon, J. H., DeLuca, N. M., Thorp, E., Liyanapatirana, C., Allen, L., Andrew, J., & Kondash, A. J. (2025). Hold my beer: The linkage between municipal water and brewing location on PFAS in popular beverages. *Environmental Science & Technology, 59*(17), 8368–8379. https://doi.org/10.1021/acs.est.4c11265

Ssebugere, P., Sillanpää, M., Matovu, H., Wang, Z., Schramm, K.-W., Omwoma, S., Wanasolo, W., Ngeno, E. C., & Odongo, S. (2020). Environmental levels and human body burdens of per-and poly-fluoroalkyl substances in Africa: A critical review. *Science of the Total Environment, 739*, 139913.

Torres, F. B. M., Guida, Y., Weber, R., & Torres, J. P. M. (2022). Brazilian overview of per- and polyfluoroalkyl substances listed as persistent organic pollutants in the Stockholm convention. *Chemosphere, 291*, 132674.

US EPA. (2024a). *Per- and Polyfluoroalkyl Substances (PFAS) Final PFAS National Primary Drinking Water Regulation.* US EPA. https://www.epa.gov/sdwa/and-polyfluoroalkyl -substances-pfas
US EPA. (2024b). *PFAS Strategic Roadmap: EPA's Commitments to Action 2021–2024.* US EPA. https://www.epa.gov/pfas/pfas-strategic-roadmap-epas-commitments-action -2021-2024
Wang, X., Zhang, H., He, X., Liu, J., Yao, Z., Zhao, H., Yu, D., Liu, B., Liu, T., & Zhao, W. (2023). Contamination of per-and polyfluoroalkyl substances in the water source from a typical agricultural area in North China. *Frontiers in Environmental Science, 10,* 1071134.

5 Analytical Methods and Monitoring for Perfluoroalkyl and Polyfluoroalkyl Substances

5.1 INTRODUCTION

In the early 2000s, the first studies demonstrating the pervasiveness of PFASs in animals were published. One such study was conducted by Giesy & Kannan (2001), which thoroughly measured the prevalence of PFASs in an animal species variety throughout the Pacific Ocean and several continents. Both densely and poorly inhabited areas were represented among the places chosen. According to the research, animals from regions with a larger concentration of humans had higher levels of PFASs overall. Researchers also discovered that PFAS concentrations were higher in predators that ate fish than in their prey, providing evidence that these contaminants may bioaccumulate and biomagnify (McCarthy et al., 2017; Boisvert et al., 2019; Miranda et al., 2022). According to further studies, these pollutants can be present in living organisms from areas with little to no human activity (Sunderland et al., 2019). Lately, more focus has been placed on examining the presence, bioaccumulation, biomagnification, and persistence of these substances and the consequences of exposure to them (Ahrens & Bundschuh, 2014; Sunderland et al., 2019; Perovani et al., 2023). Much work has been done to create sensitive and trustworthy analytical techniques for their proper determination as well as to evaluate their effect and occurrence (Burkhard & Votava, 2023; Domingo & Nadal, 2019; Ghisi et al., 2019; Jia et al., 2022; Lewis et al., 2022; Nakayama et al., 2019).

So far, a wide range of methods has been employed to detect legacy PFASs (compounds (Glüge et al., 2020; Dawson et al., 2023; Redmon et al., 2025), however, with the identification of novel PFASs in the environment, new approaches must be created to provide a sensitive and efficient way to analyze both kinds of substances (Gao et al., 2020). The fact that many of the newly identified PFASs lack standards, which makes it challenging to measure them, and that their potential ecotoxicity is yet unclear, makes this a complex problem (Gao et al., 2020; Pan et al., 2020). Recent studies have, therefore, made advancements in the determination of PFAS across

DOI: 10.1201/9781003625537-5

diverse matrices types (Glüge et al., 2020; Ogunbiyi et al., 2023; Rehman et al., 2023; Zarębska & Bajkacz, 2023; Dawson et al., 2023; Redmon et al., 2025).

5.2 SAMPLE PREPARATION

An essential first step in every analytical process is sample preparation. There are many different elements to consider when it comes to PFAS, which explains the range of ways that have been effectively implemented. Perfluoroalkyl substances (PFAS) have been found in a matrices range, containing drinking water and environmental samples, due to the importance of the possible dangers associated with these compounds (Deng et al., 2018; Janda et al., 2019; Jurikova et al., 2022; Kaboré et al., 2018; Li et al., 2022; Liu et al., 2020; Olomukoro et al., 2021; Scher et al., 2018; Vughs et al., 2019; Xian et al., 2020): surface water (Awchi et al., 2022; Ayala-Cabrera et al., 2020; Barreca et al., 2018; Janda et al., 2019; Li et al., 2022; Olomukoro et al., 2021; Song et al., 2018; Vughs et al., 2019; Wang et al., 2018a; Wang et al., 2018b; Woudneh et al., 2019; Xian et al., 2020), ground water (Barreca et al., 2018; Janda et al., 2019; Liu et al., 2020; Vughs et al., 2019), wastewater (Wang et al., 2018; Woudneh et al., 2019; Xian et al., 2020), soils (Munoz et al., 2018; Scher et al., 2018; Song et al., 2018), beverages (Xian et al., 2020), food samples (i.e. shellfish) (Abafe et al., 2021), milk and dairy products (Macheka et al., 2021; Hill et al., 2022; Sznajder-Katarzyńska et al., 2019), fruits and vegetables (Li et al., 2022; Meng et al., 2022; Rawn et al., 2022; Scher et al., 2018), edible oils (Fan et al., 2021), eggs (Chiumiento et al., 2023), prepared meals (Rawn et al., 2022), etc.), biological samples (i.e. serum) (Gao et al., 2018; Salihović et al., 2020), liver (Androulakakis et al., 2022; Barola et al., 2020; Liu et al., 2018a; 2018b; 2018c), animal tissues (Androulakakis et al., 2022; Cui et al., 2018), urine, nails, and hair (Wang et al., 2018a), small particles suspended in water (Wang et al., 2018b; 2018c), air (N. Yu et al., 2018), textiles (Drage et al., 2023), food contact materials (Vavrouš et al., 2019; Zabaleta et al., 2020; Miralles et al., 2023) etc. The vast range of characteristics found in the matrices under analysis suggests that there is general worry about the chemicals' existence in many environmental niches as well as in living things.

It is common to see several extraction procedures being utilized during the identification of these analytes due to the varying nature of the investigated materials. Given that PFASs may exist at even trace levels, these methods ought to be able to provide a target analyte with high preconcentration. SPE (solid-phase extraction), for example (Awchi et al., 2022; Barola et al., 2020; Barreca et al., 2018; Chiumiento et al., 2023; Deng et al., 2018; Janda et al., 2019; Jurikova et al., 2022; Li et al., 2022; Liu et al., 2020; Meng et al., 2022; Munoz et al., 2018; Rawn et al., 2022; X. Song et al., 2018; Vughs et al., 2019; Wang et al., 2018b), utilizing various sorbents, generally, polymeric weak anionic exchange (ANX) sorbents) (Awchi et al., 2022; Barola et al., 2020; Barreca et al., 2018; Chiumiento et al., 2023; Janda et al., 2019; Jurikova et al., 2022; Li et al., 2022; S. Liu et al., 2020; Meng et al., 2022; Munoz et al., 2018; Rawn et al., 2022; Song et al., 2018; Vughs et al., 2019; Wang et al., 2018b), although natural sorbents (Deng et al., 2018) and combinations of different sorbents (Wang et al., 2018a; 2018c) has also been found to be used to identify PFAS in water samples according to a variety of publications. The nature of these chemicals explains why ANX sorbents are frequently used in SPE methods, as demonstrated by the data given. Since PFASs are known to have a variable length carbon chain as well as a terminal functional group, often a sulfonic or carboxylic acid, some of these substances have a robust acidic character. Ionic exchange sorbents are a frequent choice for extracting these chemicals since many of these molecules are present in an anionic form (Militao et al., 2021). However, given the variety of samples examined and the fact that alternative approaches are equally feasible, rapid solvent extraction (Androulakakis et al., 2022), ultrasound-assisted extraction (Drage et al., 2023; Ferrario et al., 2021; Miralles et al., 2023; Munoz et al., 2018; Scher et al., 2018; Wang et al., 2018a, 2018c; Yu et al., 2018), solid-phase microextraction (Ayala-Cabrera et al., 2020; Olomukoro et al., 2021), solid-liquid extraction (Abafe et al., 2021; Gao et al., 2018; Liu et al., 2018b; Timshina et al., 2021; Wang et al., 2018a, 2018c; Zhou et al., 2019), dispersive solid-phase microextraction (Xian et al., 2020), liquid-liquid extraction (Hill et al., 2022; Macheka et al., 2021; Sznajder-Katarzyńska et al., 2019; Vavrouš et al., 2019), dispersive liquid-liquid microextraction (Fan et al., 2021), or a protein precipitation approach (Salihović et al., 2020) have also been employed.

It is not surprising that more publications are using newly synthesized materials in their extraction processes since they provide several benefits (such as reduced sorbent/solvent consumption, increased enrichment factors, etc.). Numerous instances of their effective application may be found in the scientific literature. Among these substances are MOFs, or metal-organic frameworks (Cheng et al., 2020; Li et al., 2021), ionic liquids (Dong et al., 2021; Shahabi Nejad et al., 2021), molecularly imprinted polymers (MIPs) (Ren et al., 2023; Zou et al., 2022), deep eutectic solvents (Fan et al., 2021), and carbonaceous material (Becanova et al., 2021; Song et al., 2023) are the most important types of new compounds that can be underlined.

5.3 SEPARATION USING CHROMATOGRAPHY AND ELECTROPHORESIS

Regarding the favoured chromatographic (CGP) methods employed in the identification of these substances, it is evident that LC, particularly in its ultra-high-performance LC (UHPLC) form, is commonly paired with MS (Miralles et al., 2023) or tandem MS detectors (MS/MS) (Vughs et al., 2019; Zhou et al., 2019; Janda et al., 2019; Woudneh et al., 2019; Xian et al., 2020; Liu et al., 2020; Barola et al., 2020; Fan et al., 2021; Macheka et al., 2021; Olomukoro et al., 2021; Timshina et al., 2021; Awchi et al., 2022; Hill et al., 2022; Jurikova et al., 2022; Li et al., 2022; Li et al., 2022; Meng et al., 2022; Rawn et al., 2022; Chiumiento et al., 2023; Drage et al., 2023).

This makes sense given the several benefits this method offers. Additionally, high-resolution mass spectrometry (HRMS) has demonstrated significant promise in PFAS research, supporting the identification of novel PFASs. Additionally, non-targeted analysis using various sample types has allowed remediation strategies to be planned before emerging PFASs become an international concern, which is exceedingly challenging with traditional targeted analysis due to the continuous influx of new pollutants that make it challenging to get standards for each of the pollutants (Liu et al., 2019a, 2019b). However, it is important to remember that a lot of these identifications are preliminary and that further focused research is required (Liu et al., 2019a, 2019b). Recapping using the CGP methods, it's important to remember that HPLC (Deng et al., 2018; Ferrario et al., 2021; Liu et al., 2018b; Song et al., 2018; Wang et al., 2018a, 2018c; Yu et al., 2018), micro-LC (Sznajder-Katarzyńska et al., 2019), or GC with MS or MS/MS sensors or detectors (Ayala-Cabrera et al., 2020) have also been used. The kind of PFASs is often a determining factor in the choice of CGP technology; ionic-neutral PFASs are studied by LC, whilst volatile/semi-volatile and neutral or unbiased ones are evaluated by GC. Regarding the kinds of columns used in the CGP separation of these substances using LC, C18 columns are used (Barreca et al., 2018; Deng et al., 2018; Scher et al., 2018; Song et al., 2018; Zhou et al., 2019; Abafe et al., 2021; Androulakakis et al., 2022; Awchi et al., 2022; Barola et al., 2020; Drage et al., 2023; Fan et al., 2021; Ferrario et al., 2021; Hill et al., 2022; Jurikova et al., 2022; Li et al., 2022; Liu et al., 2020; Macheka et al., 2021; Meng et al., 2022; Miralles et al., 2023; Olomukoro et al., 2021; Rawn et al., 2022; Salihović et al., 2020; Timshina et al., 2021; Vughs et al., 2019; Wang et al., 2018a, 2018b; 2018c; Xian et al., 2020; Chiumiento et al., 2023) were the recommended choices in the majority of the reviewed literature, with C8 columns being an exception (Scher et al., 2018), phenyl-hexyl (Vavrouš et al., 2019) and perfluorinated phenyl (Li et al., 2022) were also applied. These columns were constructed with a range of inner diameters (2.10 to 3.00 mm), particle dimensions (1.70 to 5.00 μm), and lengths (ranging from 50 to 150 mm). In reference to the GC example, the most often chosen option was 1.4 μm; 60 m x 0.25 mm; 6% cyanopropylphenyl 94% dimethyl polysiloxane (Ayala-Cabrera et al., 2020). When it comes to LC, the majority of publications have used gradient mode, often beginning with a larger proportion of an aqueous mobile phase (75–100%) (Janda et al., 2019; Vavrouš et al., 2019; Woudneh et al., 2019; Salihović et al., 2020; Abafe et al., 2021; Androulakakis et al., 2022; Awchi et al., 2022; Barola et al., 2020; Chiumiento et al., 2023; Drage et al., 2023; Fan et al., 2021; Ferrario et al., 2021; Li et al., 2022; Liu et al., 2020; Macheka et al., 2021; Meng et al., 2022; Miralles et al., 2023; Olomukoro et al., 2021; Rawn et al., 2022; Xian et al., 2020) and increasingly transitioned to an advanced degree of organic stage. With reference to the categories of organic solvents employed, methanol (MeOH) (Abafe et al., 2021; Androulakakis et al., 2022; Barola et al., 2020; Drage et al., 2023; Fan et al., 2021; Ferrario et al., 2021; Hill et al., 2022; Jurikova et al., 2022; Li et al., 2022; Macheka et al., 2021; Miralles et al., 2023; Olomukoro et al., 2021; Salihović et al., 2020; Timshina et al., 2021; Xian et al., 2020) was the most common afterwards acetonitrile (ACN) (Awchi et al., 2022; Sznajder-Katarzyńska et al., 2019; Woudneh et al., 2019), even though in some cases mixtures of MeOH: water (Barreca et al., 2018; Meng et al., 2022), or MeOH: ACN (Chiumiento et al., 2023; Liu et al., 2020; Rawn et al., 2022) were also put to work. Moreover, it has been customary to employ additives to enhance the compounds' separation and determination, like ammonium format (Drage et al., 2023; Janda et al., 2019; Li et al., 2022; Olomukoro et al., 2021; Rawn et al., 2022; Timshina et al., 2021) and NH_4OAc (Deng et al., 2018; Abafe et al., 2021; Androulakakis et al., 2022; Awchi et al., 2022; Chiumiento et al., 2023; Fan et al., 2021; Ferrario et al., 2021; Hill et al., 2022; Jurikova et al., 2022; Liu et al., 2020; Macheka et al., 2021; Meng et al., 2022; Miralles et al., 2023) being the most significant (as previous papers have emphasized) (Mullin et al., 2019), but formic acid and other chemicals were also used (Sznajder-Katarzyńska et al., 2019; Vughs et al., 2019), 1-methyl pyridine (Liu et al., 2020), and ammonium hydroxide (Liu et al., 2020). The goal of choosing the additive is to improve the ionization of the analytes and achieve stable spray dynamics in electrospray ionization (Mullin et al., 2019). The oven temperatures in these studies ranged from 25 to 50 °C, while the flow rates from 0.05 to 0.8 mL/min.

Although there are many articles on PFAS analysis, this review article has collated and commented on a few sample and current instances to provide readers with an overview of the subject. For instance, Zhou et al. (2019), by a UHPLC-MS/MS system, 20 PFAS were identified in vegetables. Solid-liquid extraction (SLE) was used to extract the material, 15.00 g of sample were weighed, and ACN (15.00 mL containing 1% formic acid) was added. With this technique, little relative standard deviations (RSDs, < 14.80%) and high recovery values (55–119%) were attained together with low limits of quantification (LOQs) of 3.00×10^{-3} to 3.40×10^{-3} µg/kg. All of the samples in this investigation had PFOA verified to be present.

Androulakakis et al. (2022) used UHPLC-MS/MS to determine the analytes and examined 56 PFASs in the muscle tissue of fish and the livers of apex predators. The contents of the samples (0.2 g of liver and 1 g of muscle sample) were extracted using accelerated solvent extraction (ASE). Following rotary evaporation, dilution, n-hexane defatting, and a final SPE clean-up process, the samples were eluted using a 50:50 v/v solution that included 1.7% formic acid (v/v) and 2% ammonia hydroxide (v/v). Low quantification (LOQs) and detection (LODs) of 0.02–1.25 µg/kg and 0.05–3.79 µg/kg (wet weight-ww), respectively, were attained using this approach. RSDs were less than 20%, and recovery values varied from 40 to 137%. The findings showed that PFOS, PFCAs, and perfluoropropyl iodides (PFPIs) were present in all specimens, with predators and specimens from urban and agricultural settings having the highest quantities.

Wang et al. (2018a, 2018b, 2018c) finally assessed whether 23 PFASs were found in river water and suspended particles. Two distinct approaches were used to obtain extraction. ACN and 1% formic acid were used in the SLE extraction of suspended particles. On the other hand, SPE was used to extract analytes from river waters utilizing a weak ANX polymer as the sorbent as well as an elution solvent of 5 mM ammonium acetate. All things considered, the analytes were identified by UHPLC-MS/MS. It was discovered that the limits of detection (LODs) for water and suspended particles were low, ranging from 0.0005 to 0.010 µg/L and 0.5 to 10 µg/kg, respectively. The recovery results were good, with RSDs continuously below 9.6% and a range of 69 to 118%. In real samples, levels of specific perfluoroalkyl substances (PFASs) ranged from 2.63 to 6.04 µg/kg for suspended composites and 4.3 to 63.8 µg/L for water from the river.

Even while ionized and polar PFASs have been identified utilizing capillary electrophoresis (CE) in addition to CGP techniques, the LODs obtained are often not as small as those achieved utilizing CGP methods (Al Amin et al., 2020) and fewer chemicals have been investigated. It is possible that more journals will use this method if improvements are achieved, even if there haven't been many articles in recent years that use CE for PFAS detection (Höcker et al., 2020). Azab et al. (2020) used multisegmented injection-non-aqueous CE-MS/MS (MSI-NACE-MS/MS) to determine the existence of PFOS and PFOA in serum samples. The electrolyte mixture used as a background consisted of 35 mM NH_4OAc in ACN (70% v/v), isopropanol (5% v/v), and CH_3OH (15% v/v) at pH 9.5. The LODs and LOQs for the technique ranged from 8.28 to 12.5 µg/L and 42.78 to 58.50 µg/L, respectively, with low RSDs (< 20%). Between 97% and 109% was the range of the mean recovery values. In another paper using CE-MS, Höcker et al. (2020) found two PFASs (perfluorooctane sulfonamide, or FOSA) and other contaminants in consumption water. The background electrolyte used was 10% isopropanol and 10% acetic acid. The ability of this technique to identify contaminants in drinking water without preparing a sample is very intriguing. With the suggested approach, low LOQs (0.04–0.4 µg/L) were frequently attained. Finally, it should be mentioned, that the lack of current publications indicates that the employment of CE methods is far more usual than more commonly used methods, such as LC or GC techniques.

5.4   SENSORS

The majority of research publications on the detection of PFASs employ CGP methods. As time has progressed, more publications have employed other sensors to detect these compounds directly. This is due to the fact that these devices are easy to use, reasonably priced, and provide on-site testing, which enables prompt screening and prompt action if required. Despite their benefits, sensors have drawbacks which need to be put into consideration, together with the sensitivity, selectivity, sample treatment/management/preconcentration, as well as mobility. In contrast to other methods like CGP processes, most of the research that has looked at PFASs has focused on finding one or a limited number of compounds (Menger et al., 2021; Rehman et al., 2023). Over the years, there have been several assessments on the use of sensors in PFAS detection (Wang et al., 2021; Rehman et al., 2023).

Sensors come in a wide range of varieties, including smartphone-based sensors, optical fibers or immunosensors, colorimetric, fluorescence, electrochemical, and ion-selective electrodes, among others (Menger et al., 2021; Rehman et al., 2023; Rodriguez et al., 2020). Coated nanoparticles (NPs), dyes, quantum dots, covalent organic frameworks (COFs), MOFs, MIPs, and carbonaceous composites are among the several detecting methods used in their preparation.

For example, in the research study by Li et al. (2019), a "turn-off" fluorescent sensor was utilized to analyze tap water and employ COF-functionalized up-conversion NPs as a fluorescence probe for PFOS in the packaging of food. In this process, water samples were filtered over a 0.22 µm nylon membrane before being dried in a rotary evaporator. Fluorescence was measured after the extract was redissolved in 20 millilitres of dimethylformamide (DMF) and 500 microlitres of a solution comprising a probe were mixed with 500 microlitres of the sample that redissolved. Regarding food packaging, samples were broken up into small pieces and 2 g were extracted using ultrasound-assisted extraction (UAE) and 20 mL of MeOH for 25 minutes at 25 °C. The extraction process was the same as that used for water samples. An LOD of 0.0621 μg/L and excellent recovery values ranging from 103% to 108% were obtained using the suggested approach. An additional publication by Fang et al. (2018) used a portable sensor based on a smartphone to measure PFOA and PFOS in tap and groundwater colorimetrically using methylene blue. Water samples were boiled for one to two minutes and then allowed to cool to room temperature before undergoing SPE to confiscate any intrusive moieties. After that, elution was performed using 1 mL of MeOH. Following the drying of the solvent, the extract was redissolved in 10 millilitres of Milli-Q water. With a low limit of detection (0.5 μg/L), the method detected PFOA throughout a concentration ($10–1.00×10^3$ μg/L). Also, Karimian et al. (2018) used the electrochemical technique of differential pulse voltammetry to develop a poly(o-phenylenediamine) MIP for PFOS detection in tap, bottled, and distilled water. The devised technique has a very low limit of detection (20 ng/L), a concentration of 50–750 ng/L, and 82–110% significant recovery rates.

Sensor-based techniques offer a practical and cost-effective way to quickly identify PFASs in samples. However, because determining PFAS is complicated, more research is needed, particularly for developing innovative composites as well as the detection techniques in addition to several pertinent evaluations (technique performance, inter-laboratory validation, toughness, etc.), before these techniques can be widely used at the industry and regulatory agency levels (Menger et al., 2021; Rehman et al., 2023).

5.5 ANALYTICAL CHALLENGES

There are still a number of issues with PFAS determination that need more work. One of the main issues affecting the measurement of PFAS is the absence of standards, particularly when novel PFASs are discovered. Furthermore, the availability of isotope-labeled per- and polyfluoroalkyl substances (PFASs) is restricted, which poses a significant challenge in the extraction process development. These substances are necessary to improve quantification by reducing matrix effects on recoveries (Pan et al., 2020).

Difficulties with quality assurance and control lead to another problem. The likelihood that many of the conductions, solvents, and lab equipment used may contaminated with PFASs, resulting in cross-contamination, is the reason why PFAS analysis is so difficult (Lewis, 2022; Pan et al., 2020). Because of this, using PFAS-free solvents, tubing, and labware, along with conducting prescreening analyses and using laboratory blanks, are all crucial measures. Additionally, using cleanroom facilities, treating glassware at high temperatures (above 450 °C) before analysis, escaping the utilization of aluminium foil to cover glassware (a possible source of PFAS contamination), and cutting back on needless sample preparation steps might also assist to limit contamination (Pan et al., 2020; Winchell et al., 2021). For retaining and delaying the pollutants from the LC system as well as the spent movable phase, it could be a good idea to install a guard column or delay column before the injector in addition to all of these safety measures (Zhi & Liu, 2018; Pan et al., 2020; Genualdi et al., 2022; Srivastava et al., 2022).

The choice of suitable containers for sample collection and extraction is another crucial consideration; polypropylene or high-density polyethylene is the suggested material (Winchell et al., 2021). However, concerns still need to be made even with these materials because it has been shown that PFASs may eventually wind up being reserved by the plastic containers themselves (Pan et al., 2020; Zenobio et al., 2022).

Last but not least, the separation of PFASs may be a challenging topic that contributed to the MS detectors overwhelming usage in this sector. This is because the employment of MS transitions is essential to enabling additional separation of these chemicals (Berger et al., 2011). However, selective ion determination and sensitivity limit the ability to determine PFASs. This is demonstrated by the significant sensitivity variation between instruments, which has been linked to in-source fragmentation and dimer formation (Brase & Spink, 2020; Pan et al., 2020). The analysis of PFAS isomers is a complex topic for two additional reasons: the challenges of achieving a broad CGP separation of the PFASs via GC, (that achieves separation but involves an additional derivatization procedure that could possibly introduce new adulteration sources) and LC (which takes longer and requires specially designed columns) (Pan et al., 2020). Interestingly, new ambient ionization methods based on plasma have been introduced. Schütz et al., (2015) and Vogel et al., (2019) may aid in avoiding many of the issues already raised, permitting quick and inexpensive in situ measurement as well as screening evaluation of PFASs while reducing a number of issues that have plagued conventional ionization procedures, potentially representing a significant advancement in this field of study (Brown & Fedick, 2021; Emmons et al., 2023).

5.6 PFAS OCCURRENCE IN REAL SAMPLES

Over the years, a considerable number of evaluations have assessed the existence of PFASs in various matrices because of their impacts on the environment and biology. These papers contain a variety of matrices, including drinking water (Li et al., 2020), ground (Wang et al., 2022; Xu et al., 2021), surface (Wang et al., 2022), seawater (Li et al., 2020), wastewater (Kurwadkar et al., 2022), sediments (Li et al., 2020), food (Pasecnaja et al., 2022), wildlife (Chen et al., 2021; De Silva et al., 2021; Death et al., 2021), etc., have been thoroughly examined. Furthermore, the evaluation of these pollutants' origins and/or their dangers has been evaluated on several occasions (Death et al., 2021; Garg et al., 2020; Wang et al., 2022). Numerous PFASs were detected in surface and groundwater samples, with PFOAs and PFOSs being frequently detected (Xu et al., 2021). PFAS-related businesses, wastewater treatment facilities' water discharges, landfills (via PFAS leachates), and other upstream sources all have an impact on the prevalence of PFASs in various kinds of water (Podder et al., 2021). This topic has also been the subject of several reviews. For example, Domingo & Nadal (2019) provided a comprehensive investigation that assessed the PFASs present in drinking water on every continent. This article illustrated the different patterns in the prevalence of PFAS: According to Domingo and Nadal (2019), long-chain PFASs (hydrophobic) are more likely to be retained in sediments, whereas short-chain PFASs (hydrophilic) are more common in ambient waters. According to the results of the publications gathered, there is now not a high ecotoxicological danger associated with the levels of PFASs in drinking water, it was noted in this study. Since drinking water treatment facilities get cleaner influents with lower PFAS concentrations than wastewater treatment plants, these results may be explained by the drinking water treatment plants' superior PFAS removal capabilities (Wang et al., 2022).

According to studies, PFOSs were likewise thought to be among the most prevalent PFASs in dietary samples. Pasecnaja et al. (2022) featured a wide range of culinary samples—fish, seafood, meat, dairy goods, fruits, etc.—that were accessible throughout Europe. Additionally, an assessment of how these pollutants are ingested by humans was conducted, revealing that in a number of nations, dietary exposure surpassed weekly doses that are considered acceptable. Bioaccumulation has also been a highly significant topic that has been extensively investigated (Conder et al., 2008; Gagliano et al., 2020; Lesmeister et al., 2021; Lewis et al., 2022). In their assessment of PFAS accumulation in agricultural plants, for instance, Lesmeister et al. (2021) revealed that the main source of PFAS absorption is contaminated soils. According to this research, chain length in per- and polyfluoroalkyl substances (PFASs) seems to have a better effect on transfer than functional head group modifications. In another research by Lewis et al. (2022), they assessed the procedures of exposure and bioaccumulation in freshwater settings. This review indicated that species near sediments are more likely to be exposed to more PFAS amounts, particularly PFASs with longer chains. This is an intriguing finding. Additionally, it underlined how important it is to consider trophic levels when reporting data because it may affect the biotransformation and accumulation of precursors. Researchers also said in this study that evidence indicated a connection between increased bioaccumulation and chain length and sulfonate head groups, particularly for aquatic plants. In light of the chain length importance for PFASs, Gagliano et al. (2020) showed that longer chains are better at retaining PFASs than shorter ones, which makes it more challenging to remove them from the latter. Furthermore, it was demonstrated that the presence of organic matter might alter how these chemicals sorb, with shorter-chain PFASs being particularly adversely impacted, however, the influence of organic matter on long-chain PFASs was less noticeable. It should be mentioned that, according to several evaluations, food packaging might be one of the primary routes for PFAS contamination of food (Ramírez Carnero et al., 2021; Glenn et al., 2021; Barhoumi et al., 2022). Since they can be a possible source of exposure, it is thus very interesting to evaluate if PFASs are present in the materials that are used to make food.

Lastly, a number of papers have also covered biomagnification (McCarthy et al., 2017; Miranda et al., 2022) Nevertheless, there aren't many evaluations on this subject. Miranda et al. (2022) examined, for example, the worldwide variance in biomagnification among various aquatic compartments and the environmental parameters affecting this process. The danger of PFAS biomagnification and the need for these pollutants to be regulated were emphasized in this study. McCarthy et al. (2017) further examined a few papers that evaluated the biomagnification of PFASs, highlighting the dearth of standardized reporting and the wide range of biomagnification and bioaccumulation studies in PFASs study.

A lot of work has been put into identifying as many PFASs as possible in various sample types by journals in the scientific literature, as PFASs have demonstrated the potential to be detrimental to both individuals and the environment. . PFASs have been found in samples from all across the world including Asia, according to these investigations: China (Deng et al., 2018; Fan et al., 2021; Liu et al., 2020; Xian et al., 2020), South Korea (Seo et al., 2019), and Japan (Kaboré et al., 2018); Africa: South Africa (Abafe et al., 2021; Macheka et al., 2021), Ivory Coast (Kaboré et al., 2018), and Burkina Faso (Kaboré et al., 2018); the Americas: United States of America (Hill et al., 2022; Li et al., 2022; Meng et al., 2022; Olomukoro et al., 2021), Mexico (Kaboré et al., 2018), Canada (Kaboré et al., 2018; Rawn et al., 2022, and Chile (Kaboré et al., 2018); and Europe: Germany (Androulakakis et al., 2022; Janda et al., 2019), Ireland (Drage et al., 2023), United Kingdom (Androulakakis et al., 2022), Sweden (Salihović et al., 2020), Norway (Kaboré et al., 2018), Netherlands (Awchi et al., 2022; Vughs et al., 2019), Belgium (Vughs et al., 2019), France (Kaboré et al., 2018), Czech Republic (Jurikova et al., 2022), Italy (Barola et al., 2020; Chiumiento et al., 2023), Poland (Sznajder-Katarzyńska et al., 2019), and Spain (Ayala-Cabrera et al., 2020); or Oceania: Australia (Coggan et al., 2019; Nguyen et al., 2019).

Since PFASs were seen in each of these samples, this emphasizes how commonplace they are and how dangerous they may be. In general, PFOA was one of the most often found PFASs (Deng et al., 2018; Sznajder-Zhou et al., 2019; Janda et al., 2019; Salihović et al., 2020; Abafe et al., 2021; Macheka et al., 2021; Fan et al., 2021; Awchi et al., 2022; Jurikova et al., 2022; Li et al., 2022), perfluorohexanoic acid (PFHxA) (Scher et al., 2018; Sznajder-Katarzyńska et al., 2019; Janda et al., 2019; Abafe et al., 2021; Ferrario et al., 2021; Fan et al., 2021; Macheka et al., 2021; Androulakakis et al., 2022; Awchi et al., 2022; Jurikova et al., 2022; Li et al., 2022; Drage et al., 2023), PFOS (Abafe et al., 2021; Chiumiento et al., 2023; Drage et al., 2023; Ferrario et al., 2021; Li et al., 2022; Macheka et al., 2021; Salihović et al., 2020; Zhou et al., 2019), perfluoroheptanoic acid (PFHpA) (Abafe et al., 2021; Androulakakis et al., 2022; Awchi et al., 2022; Chiumiento et al., 2023; Drage et al., 2023; Ferrario et al., 2021; Janda et al., 2019; Jurikova et al., 2022; Li et al., 2022; Macheka et al., 2021; Sznajder-Katarzyńska et al., 2019), PFHxS (Kaboré et al., 2018; Deng et al., 2018; Woudneh et al., 2019; Salihović et al., 2020; Macheka et al., 2021; Abafe et al., 2021; Ferrario et al., 2021; Li et al., 2022; Chiumiento et al., 2023; Drage et al., 2023), perfluorodecanoic acid (PFDA) (Sznajder-Katarzyńska et al., 2019; Salihović et al., 2020; Macheka et al., 2021; Fan et al., 2021; Ferrario et al., 2021; Jurikova et al., 2022; Li et al., 2022; Awchi et al., 2022; Chiumiento et al., 2023; Drage et al., 2023), perfluorobutanoic acid (PFBA) (Abafe et al., 2021; Androulakakis et al., 2022; Fan et al., 2021; Ferrario et al., 2021; Janda et al., 2019; Kaboré et al., 2018; Li et al., 2022; Macheka et al., 2021; Woudneh et al., 2019), PFNA (Abafe et al., 2021; Awchi et al., 2022; Chiumiento et al., 2023; Ferrario et al., 2021; Jurikova et al., 2022; Li et al., 2022; Macheka et al., 2021) perfluoropentanoic acid (PFPeA) (Kaboré et al., 2018; Zhou et al., 2019; Janda et al., 2019; Abafe et al., 2021; Macheka et al., 2021; Ferrario et al., 2021; Androulakakis et al., 2022; Li et al., 2022), and perfluorobutanesulfonic acid (PFBS) (Abafe et al., 2021; Drage et al., 2023; Ferrario et al., 2021; Li et al., 2022; Macheka et al., 2021; Woudneh et al., 2019; Xian et al., 2020), although other contaminants were also identified: perfluorododecanoic acid (PFDoA) (Androulakakis et al., 2022; Drage et al., 2023; Jurikova et al., 2022; Li et al., 2022; Macheka et al., 2021; Woudneh et al., 2019), perfluoroundecanoic acid (PFUdA) (Drage et al., 2023; Jurikova et al., 2022; Li et al., 2022; Macheka et al., 2021; Xian et al., 2020), perfluorotetradecenoic acid (PFTeDA) (Abafe et al., 2021; Androulakakis et al., 2022; Kaboré et al., 2018; Li et al., 2022; Macheka et al., 2021), perfluorodecanesulfonic acid (PFDS) (Abafe et al., 2021; Drage et al., 2023; Kaboré et al., 2018; Li et al., 2022; Macheka et al., 2021; Woudneh et al., 2019), and perfluoroheptanesulfonic acid (PFHpS) (Androulakakis et al., 2022; Drage et al., 2023; Li et al., 2022; Woudneh et al., 2019).

Keeping the recent five years' research publications in focus, there are a few that are noteworthy. As an illustration, in the work of Macheka et al. (2021), in South African shellfish farms, 15 PFASs were investigated. The dispersive SPE-UHPLC-MS/MS approach that was suggested allowed the researchers to get high recoveries (77–119%, RSD < 27%) and low limit of quantification (LOQs) for each of the analytes (0.005–0.050 µg/kg). After sample analysis, a number of PFASs were found in concentrations between 0.12 and 6.43 µg/kg, with PFPeA making a total of about 85%. "PFBA, PFPeA, PFBS, PFHxA, PFHpA, PFHxS, PFOA, PFNA, PFOS, PFDS, and PFTeDA" were among these PFASs. Overall, the results indicated that there was no risk of PFAS exposure to health from eating shellfish.

Miralles et al. (2023) used UHPLC-MS to analyze 21 PFASs in cardboard and paper products that come into contact with food. The extraction process used a green analytical approach in which UAE used a 40:60 v/v combination of ACN and water, followed by centrifugation. Generally, the technique demonstrated low RSDs (< 20%) and strong recoveries (74–115%) with low LOQs (1.7–10 µg/kg). Lastly, analysis of the 16 samples revealed that all of them (with levels below the LOQ) conformed to the most recent EU requirements.

Alternative publication by Li et al. (2022) among surface and tap water samples from the USA, 30 PFASs were found. UHPLC-MS/MS was developed to identify the analyte, and traditional SPE was employed to extract using a polymeric weak-ANX sorbent. The method's recoveries ranged from 65% to 128%, while its LODs were from 0.01 to 1.99 ng/L. Surface water had 28 PFASs, whereas tap water contained 25. The concentration of PFHxS ranged from 0.004 to 0.116 µg/L in surface water and from 0.001 to 0.242 µg/L in tap water, respectively. "PFBA, PFBS, PFHxS, PFPeA, PFHxA, 4:2 fluorotelomer sulfonic acid (4:2 FTS), PFHpA, PFHpS, hexafluoropropylene oxide dimer acid (GenX), PFOA, FOSA, PFOS, 6:2 fluorotelomer sulfonic acid (6:2 FTS), PFNA, N-methylperfluorooctanesulfonamidoacetic acid (N-MeFOSAA), and PFBA" were among the PFASs found in tap water, The compounds found in surface water included "PFBA, FBSA, PFPeS, PFHxS, PFBS, PFPeA, PFHxA, PFUdA, PFTeDA, N-ethylperfluorooctanesulfonamidoacetic acid (N-EtFOSAA), PFDoA, perfluorotridecanoic acid (PFTrDA), and 8:2 fluorotelomer sulfonic acid (8:2 FTS), N-MeFOSAA, PFNA, FOSA, GenX, PFDA, N-EtFOSAA,

PFNS, 8:2 FTS, PFDoA, PFUdA, PFTrDA, 4:2 FTS, PFDS, PFHpA, PFHpS, PFOS, PFOA". Researchers found that PFASs in tap water had a high hazard or threat index, indicating that there may be a significant danger to human health from exposure to PFASs in tap water. This outcome differs from the conclusions of previous publications, including those published by Domingo & Nadal (2019). This emphasizes the necessity of standardizing analytical techniques and the significance of further PFAS analysis research.

Lastly, Chiumiento and associates (2023) discovered that 18 PFASs were found in samples of conventional and organic eggs by Italian researchers utilizing SPE and a polymeric weak ANX sorbent. The analytes were identified employing UHPLC-MS/MS. The proposed technique showed low LOQs (0.05 µg/kg) and LODs (0.005–0.036 µg/kg), together with low RSDs (< 21%) and excellent recoveries (81–128%). In actual samples, at least six PFASs were detected, with concentrations ranging from 0.006 to 0.042 µg/kg: PFDA, PFHpA, PFHxS, PFNA, PFDoA, perfluorodo-decane sulfonic acid, and PFOS, or PFDoDS. Researchers found that there were no appreciable variations between conventional and organic eggs and that the overall concentration of PFASs in eggs was low.

As previously mentioned, a large number of PFASs have been investigated world-wide in a variety of samples from diverse geographic regions. In general, PFASs were found in most samples, but in varying amounts. PFOA, PFOS, PFNA, and PFHxS have a weekly tolerated intake (TWI) of 4.4 ng/kg, and the overall limit for all PFASs in drinking water is 0.5 µg/L, according to the European Chemical Agency (ECHA) (ECHA, 2023). However, as the many PFASs found have demonstrated, the restrictions in place should be revised. The existence of so many distinct PFAS types in samples should alert scientists, as it is a clear indication of possible hazards down the road. Furthermore, since drinking water has been the only subject of regulations up until now, more work needs to be done to establish a single set of standards for evaluating other samples. The fact that PFASs have originated in a variety of samples, as was previously indicated, emphasizes the pervasive danger that these substances might present, especially when bioaccumulation is taken into consideration.

5.7   CONCLUSION

Our capacity to identify and track PFAS in a variety of biological and environmental matrices has greatly increased because of developments in analytical techniques. Complex sample preparation, matrix interferences, and the identification of new PFAS chemicals are still problems, nevertheless. In order to improve the precision and comparability of PFAS monitoring data, this chapter emphasizes the significance of creating standardized and reasonably priced analytical procedures. To overcome these obstacles, future initiatives should concentrate on combining cutting-edge sensor technologies with high-throughput analytical systems. Researchers and policymakers may get a better understanding of the prevalence, distribution, and dangers of PFAS by improving analytical techniques, which will ultimately lead to more efficient management and remediation plans.

REFERENCES

Abafe, O. A., Macheka, L. R., Abafe, O. T., & Chokwe, T. B. (2021). Concentrations and human exposure assessment of per and polyfluoroalkyl substances in farmed marine shellfish in South Africa. *Chemosphere, 281*, 130985.

Ahrens, L., & Bundschuh, M. (2014). Fate and effects of poly- and perfluoroalkyl substances in the aquatic environment: A review. *Environmental Toxicology and Chemistry, 33*(9), 1921–1929.

Al Amin, M., Sobhani, Z., Liu, Y., Dharmaraja, R., Chadalavada, S., Naidu, R., Chalker, J. M., & Fang, C. (2020). Recent advances in the analysis of per-and polyfluoroalkyl substances (PFAS)—A review. *Environmental Technology & Innovation, 19*, 100879.

Androulakakis, A., Alygizakis, N., Gkotsis, G., Nika, M.-C., Nikolopoulou, V., Bizani, E., Chadwick, E., Cincinelli, A., Claßen, D., & Danielsson, S. (2022). Determination of 56 per-and polyfluoroalkyl substances in top predators and their prey from Northern Europe by LC-MS/MS. *Chemosphere, 287*, 131775.

Awchi, M., Gebbink, W. A., Berendsen, B. J. A., Benskin, J. P., & van Leeuwen, S. P. J. (2022). Development, validation, and application of a new method for the quantitative determination of monohydrogen-substituted perfluoroalkyl carboxylic acids (H–PFCAs) in surface water. *Chemosphere, 287*, 132143.

Ayala-Cabrera, J. F., Contreras-Llin, A., Moyano, E., & Santos, F. J. (2020). A novel methodology for the determination of neutral perfluoroalkyl and polyfluoroalkyl substances in water by gas chromatography-atmospheric pressure photoionisation-high resolution mass spectrometry. *Analytica Chimica Acta, 1100*, 97–106.

Azab, S., Hum, R., & Britz-McKibbin, P. (2020). Rapid biomonitoring of perfluoroalkyl substance exposures in serum by multisegment injection-nonaqueous capillary electrophoresis-tandem mass spectrometry. *Analytical Science Advances, 1*(3), 173–182.

Barhoumi, B., Sander, S. G., & Tolosa, I. (2022). A review on per-and polyfluorinated alkyl substances (PFASs) in microplastic and food-contact materials. *Environmental Research, 206*, 112595.

Barola, C., Moretti, S., Giusepponi, D., Paoletti, F., Saluti, G., Cruciani, G., Brambilla, G., & Galarini, R. (2020). A liquid chromatography-high resolution mass spectrometry method for the determination of thirty-three per-and polyfluoroalkyl substances in animal liver. *Journal of Chromatography A, 1628*, 461442.

Barreca, S., Busetto, M., Vitelli, M., Colzani, L., Clerici, L., & Dellavedova, P. (2018). Online solid-phase extraction LC-MS/MS: A rapid and valid method for the determination of perfluorinated compounds at sub ng· L−1 level in natural water. *Journal of Chemistry, 2018*, 1–9.

Becanova, J., Saleeba, Z. S. S. L., Stone, A., Robuck, A. R., Hurt, R. H., & Lohmann, R. (2021). A graphene-based hydrogel monolith with tailored surface chemistry for PFAS passive sampling. *Environmental Science: Nano, 8*(10), 2894–2907.

Berger, U., Kaiser, M. A., Kärrman, A., Barber, J. L., & Van Leeuwen, S. P. J. (2011). Recent developments in trace analysis of poly-and perfluoroalkyl substances. *Analytical and Bioanalytical Chemistry, 400*, 1625–1635.

Boisvert, G., Sonne, C., Rigét, F. F., Dietz, R., & Letcher, R. J. (2019). Bioaccumulation and biomagnification of perfluoroalkyl acids and precursors in East Greenland polar bears and their ringed seal prey. *Environmental Pollution, 252*, 1335–1343.

Brase, R. A., & Spink, D. C. (2020). Enhanced Sensitivity for the analysis of perfluoroethercarboxylic acids using LC–ESI–MS/MS: Effects of probe position, mobile phase additive, and capillary voltage. *Journal of the American Society for Mass Spectrometry, 31*(10), 2124–2132.

Brown, H. M., & Fedick, P. W. (2021). Rapid, low-cost, and in-situ analysis of per-and poly-fluoroalkyl substances in soils and sediments by ambient 3D-printed cone spray ioniza-tion mass spectrometry. *Chemosphere, 272*, 129708.

Burkhard, L. P., & Votava, L. K. (2023). Review of per-and polyfluoroalkyl substances (PFAS) bioaccumulation in earthworms. *Environmental Advances, 11*, 100335.

Chen, Y., Fu, J., Ye, T., Li, X., Gao, K., Xue, Q., Lv, J., Zhang, A., & Fu, J. (2021). Occurrence, profiles, and ecotoxicity of poly-and perfluoroalkyl substances and their alternatives in global apex predators: A critical review. *Journal of Environmental Sciences, 109*, 219–236.

Cheng, Y. H., Barpaga, D., Soltis, J. A., Shutthanandan, V., Kargupta, R., Han, K. S., McGrail, B. P., Motkuri, R. K., Basuray, S., & Chatterjee, S. (2020). Metal–organic framework-based microfluidic impedance sensor platform for ultrasensitive detection of perfluo-rooctanesulfonate. *ACS Applied Materials & Interfaces, 12*(9), 10503–10514.

Chiumiento, F., Bellocci, M., Ceci, R., D'Antonio, S., De Benedictis, A., Leva, M., Pirito, L., Rosato, R., Scarpone, R., & Scortichini, G. (2023). A new method for determining PFASs by UHPLC-HRMS (Q-Orbitrap): Application to PFAS analysis of organic and conventional eggs sold in Italy. *Food Chemistry, 401*, 134135.

Coggan, T. L., Moodie, D., Kolobaric, A., Szabo, D., Shimeta, J., Crosbie, N. D., Lee, E., Fernandes, M., & Clarke, B. O. (2019). An investigation into per-and polyfluoroalkyl substances (PFAS) in nineteen Australian wastewater treatment plants (WWTPs). *Heliyon, 5*(8), Article e02316..

Conder, J. M., Hoke, R. A., Wolf, W. de, Russell, M. H., & Buck, R. C. (2008). Are PFCAs bioaccumulative? A critical review and comparison with regulatory criteria and per-sistent lipophilic compounds. *Environmental Science & Technology, 42*(4), 995–1003.

Cui, Q., Pan, Y., Zhang, H., Sheng, N., Wang, J., Guo, Y., & Dai, J. (2018). Occurrence and tissue distribution of novel perfluoroether carboxylic and sulfonic acids and legacy per/polyfluoroalkyl substances in black-spotted frog (Pelophylax nigromaculatus). *Environmental Science & Technology, 52*(3), 982–990. https://doi.org/10.1021/acs.est.7b03662

Dawson, D. E., Lau, C., Pradeep, P., Sayre, R. R., Judson, R. S., Tornero-Velez, R., & Wambaugh, J. F. (2023). A machine learning model to estimate toxicokinetic half-lives of per-and polyfluoro-alkyl substances (PFAS) in multiple species. *Toxics, 11*(2), 98. https://doi.org/10.3390/toxics11020098

De Silva, A. O., Armitage, J. M., Bruton, T. A., Dassuncao, C., Heiger-Bernays, W., Hu, X. C., Kärrman, A., Kelly, B., Ng, C., & Robuck, A. (2021). PFAS exposure pathways for humans and wildlife: A synthesis of current knowledge and key gaps in understanding. *Environmental Toxicology and Chemistry, 40*(3), 631–657.

Death, C., Bell, C., Champness, D., Milne, C., Reichman, S., & Hagen, T. (2021). Per-and polyfluoroalkyl substances (PFAS) in livestock and game species: A review. *Science of The Total Environment, 774*, 144795.

Deng, Z.-H., Cheng, C.-G., Wang, X.-L., Shi, S.-H., Wang, M.-L., & Zhao, R.-S. (2018). Preconcentration and determination of perfluoroalkyl substances (PFASs) in water samples by bamboo charcoal-based solid-phase extraction prior to liquid chromatogra-phy–tandem mass spectrometry. *Molecules, 23*(4), 902.

Domingo, J. L., & Nadal, M. (2019). Human exposure to per-and polyfluoroalkyl sub-stances (PFAS) through drinking water: A review of the recent scientific literature. *Environmental Research, 177*, 108648.

Dong, Q., Min, X., Huo, J., & Wang, Y. (2021). Efficient sorption of perfluoroalkyl acids by ionic liquid-modified natural clay. *Chemical Engineering Journal Advances, 7*, 100135.

Drage, D. S., Sharkey, M., Berresheim, H., Coggins, M., & Harrad, S. (2023). Rapid determi-nation of selected PFAS in textiles entering the waste stream. In *Toxics* (Vol. 11, Issue 1). https://doi.org/10.3390/toxics11010055

ECHA. (2023). *Per- and polyfluoroalkyl substances (PFAS)*. https://echa.europa.eu/hot-topics/perfluoroalkyl-chemicals-pfas

Emmons, R. V., Fatigante, W., Olomukoro, A. A., Musselman, B., & Gionfriddo, E. (2023). Rapid screening and quantification of PFAS enabled by SPME-DART-MS. *Journal of the American Society for Mass Spectrometry*, 34(9), 1890–1897.

Fan, C., Wang, H., Liu, Y., & Cao, X. (2021). New deep eutectic solvent based superparamagnetic nanofluid for determination of perfluoroalkyl substances in edible oils. *Talanta*, 228, 122214.

Fang, C., Zhang, X., Dong, Z., Wang, L., Megharaj, M., & Naidu, R. (2018). Smartphone app-based/portable sensor for the detection of fluoro-surfactant PFOA. *Chemosphere*, 191, 381–388.

Ferrario, C., Valsecchi, S., Lava, R., Bonato, M., & Polesello, S. (2021). Determination of perfluoroalkyl acids in different tissues of graminaceous plants. *Analytical Methods*, 13(13), 1643–1650. https://doi.org/10.1039/D0AY02226H

Gagliano, E., Sgroi, M., Falciglia, P. P., Vagliasindi, F. G. A., & Roccaro, P. (2020). Removal of poly-and perfluoroalkyl substances (PFAS) from water by adsorption: Role of PFAS chain length, effect of organic matter and challenges in adsorbent regeneration. *Water Research*, 171, 115381.

Gao, K., Chen, Y., Xue, Q., Fu, J., Fu, K., Fu, J., Zhang, A., Cai, Z., & Jiang, G. (2020). Trends and perspectives in per-and polyfluorinated alkyl substances (PFASs) determination: Faster and broader. *TrAC Trends in Analytical Chemistry*, 133, 116114.

Gao, K., Fu, J., Xue, Q., Li, Y., Liang, Y., Pan, Y., Zhang, A., & Jiang, G. (2018). An integrated method for simultaneously determining 10 classes of per-and polyfluoroalkyl substances in one drop of human serum. *Analytica Chimica Acta*, 999, 76–86.

Garg, S., Kumar, P., Mishra, V., Guijt, R., Singh, P., Dumée, L. F., & Sharma, R. S. (2020). A review on the sources, occurrence and health risks of per-/poly-fluoroalkyl substances (PFAS) arising from the manufacture and disposal of electric and electronic products. *Journal of Water Process Engineering*, 38, 101683.

Genualdi, S., Beekman, J., Carlos, K., Fisher, C. M., Young, W., DeJager, L., & Begley, T. (2022). Analysis of per-and poly-fluoroalkyl substances (PFAS) in processed foods from FDA's Total Diet Study. *Analytical and Bioanalytical Chemistry*, 414(3), 1189–1199.

Ghisi, R., Vamerali, T., & Manzetti, S. (2019). Accumulation of perfluorinated alkyl substances (PFAS) in agricultural plants: A review. *Environmental Research*, 169, 326–341.

Giesy, J. P., & Kannan, K. (2001). Global distribution of perfluorooctane sulfonate in wildlife. *Environmental Science & Technology*, 35(7), 1339–1342.

Glenn, G., Shogren, R., Jin, X., Orts, W., Hart-Cooper, W., & Olson, L. (2021). Per- and polyfluoroalkyl substances and their alternatives in paper food packaging. *Comprehensive Reviews in Food Science and Food Safety*, 20(3), 2596–2625.

Glüge, J., Scheringer, M., Cousins, I. T., DeWitt, J. C., Goldenman, G., Herzke, D., Lohmann, R., Ng, C. A., Trier, X., & Wang, Z. (2020). An overview of the uses of per-and polyfluoroalkyl substances (PFAS). *Environmental Science: Processes & Impacts*, 22(12), 2345–2373. https://doi.org/10.1039/D0EM00291G

Hill, N. I., Becanova, J., & Lohmann, R. (2022). A sensitive method for the detection of legacy and emerging per-and polyfluorinated alkyl substances (PFAS) in dairy milk. *Analytical and Bioanalytical Chemistry*, 414, 1235–1243.

Höcker, O., Bader, T., Schmidt, T. C., Schulz, W., & Neusüß, C. (2020). Enrichment-free analysis of anionic micropollutants in the sub-ppb range in drinking water by capillary electrophoresis-high resolution mass spectrometry. *Analytical and Bioanalytical Chemistry*, 412, 4857–4865.

Janda, J., Nödler, K., Brauch, H.-J., Zwiener, C., & Lange, F. T. (2019). Robust trace analysis of polar (C2-C8) perfluorinated carboxylic acids by liquid chromatography-tandem mass spectrometry: Method development and application to surface water, groundwater and drinking water. *Environmental Science and Pollution Research, 26*(8), 7326–7336.

Jia, S., Marques Dos Santos, M., Li, C., & Snyder, S. A. (2022). Recent advances in mass spectrometry analytical techniques for per-and polyfluoroalkyl substances (PFAS). *Analytical and Bioanalytical Chemistry, 414*(9), 2795–2807.

Jurikova, M., Dvorakova, D., & Pulkrabova, J. (2022). The occurrence of perfluoroalkyl substances (PFAS) in drinking water in the Czech Republic: A pilot study. *Environmental Science and Pollution Research, 29*(40), 60341–60353.

Kaboré, H. A., Duy, S. V., Munoz, G., Méité, L., Desrosiers, M., Liu, J., Sory, T. K., & Sauvé, S. (2018). Worldwide drinking water occurrence and levels of newly-identified perfluoroalkyl and polyfluoroalkyl substances. *Science of The Total Environment, 616,* 1089–1100.

Karimian, N., Stortini, A. M., Moretto, L. M., Costantino, C., Bogialli, S., & Ugo, P. (2018). Electrochemosensor for trace analysis of perfluorooctanesulfonate in water based on a molecularly imprinted poly (o-phenylenediamine) polymer. *ACS Sensors, 3*(7), 1291–1298.

Kurwadkar, S., Dane, J., Kanel, S. R., Nadagouda, M. N., Cawdrey, R. W., Ambade, B., Struckhoff, G. C., & Wilkin, R. (2022). Per-and polyfluoroalkyl substances in water and wastewater: A critical review of their global occurrence and distribution. *Science of The Total Environment, 809,* 151003.

Lesmeister, L., Lange, F. T., Breuer, J., Biegel-Engler, A., Giese, E., & Scheurer, M. (2021). Extending the knowledge about PFAS bioaccumulation factors for agricultural plants– A review. *Science of The Total Environment, 766,* 142640.

Lewis, A. J., Yun, X., Spooner, D. E., Kurz, M. J., McKenzie, E. R., & Sales, C. M. (2022). Exposure pathways and bioaccumulation of per-and polyfluoroalkyl substances in freshwater aquatic ecosystems: Key considerations. *Science of the Total Environment, 822,* 153561.

Li, F., Duan, J., Tian, S., Ji, H., Zhu, Y., Wei, Z., & Zhao, D. (2020). Short-chain per-and polyfluoroalkyl substances in aquatic systems: Occurrence, impacts and treatment. *Chemical Engineering Journal, 380,* 122506.

Li, J., Zhang, C., Yin, M., Zhang, Z., Chen, Y., Deng, Q., & Wang, S. (2019). Surfactant-sensitized covalent organic frameworks-functionalized lanthanide-doped nanocrystals: An ultrasensitive sensing platform for perfluorooctane sulfonate. *ACS Omega, 4*(14), 15947–15955.

Li, N., Song, X., Shen, P., & Zhao, C. (2022). Rapid determination of perfluoroalkyl and polyfluoroalkyl substances (PFASs) in vegetables by on-line solid-phase extraction (SPE) with ultra-high-performance liquid chromatography-tandem mass spectrometry (UHPLC-MS/MS). *Analytical Letters, 55*(14), 2227–2238.

Li, R., Alomari, S., Islamoglu, T., Farha, O. K., Fernando, S., Thagard, S. M., Holsen, T. M., & Wriedt, M. (2021). Systematic study on the removal of per-and polyfluoroalkyl substances from contaminated groundwater using metal–organic frameworks. *Environmental Science & Technology, 55*(22), 15162–15171.

Li, X., Fatowe, M., Cui, D., & Quinete, N. (2022). Assessment of per-and polyfluoroalkyl substances in Biscayne Bay surface waters and tap waters from South Florida. *Science of the Total Environment, 806,* 150393.

Liu, G., Dhana, K., Furtado, J. D., Rood, J., Zong, G., Liang, L., Qi, L., Bray, G. A., DeJonge, L., Coull, B., Grandjean, P., & Sun, Q. (2018a). Perfluoroalkyl substances and changes in body weight and resting metabolic rate in response to weight-loss diets: A prospective study. *PLOS Medicine, 15*(2), e1002502. https://doi.org/10.1371/journal.pmed.1002502

Liu, H.-S., Wen, L.-L., Chu, P.-L., & Lin, C.-Y. (2018c). Association among total serum isomers of perfluorinated chemicals, glucose homeostasis, lipid profiles, serum protein and metabolic syndrome in adults: NHANES, 2013–2014. *Environmental Pollution, 232*, 73–79. https://doi.org/https://doi.org/10.1016/j.envpol.2017.09.019

Liu, S., Junaid, M., Zhong, W., Zhu, Y., & Xu, N. (2020). A sensitive method for simultaneous determination of 12 classes of per-and polyfluoroalkyl substances (PFASs) in groundwater by ultrahigh performance liquid chromatography coupled with quadrupole orbitrap high resolution mass spectrometry. *Chemosphere, 251*, 126327.

Liu, Y., D'Agostino, L. A., Qu, G., Jiang, G., & Martin, J. W. (2019a). High-resolution mass spectrometry (HRMS) methods for nontarget discovery and characterization of poly- and per-fluoroalkyl substances (PFASs) in environmental and human samples. *TrAC Trends in Analytical Chemistry, 121*, 115420.

Liu, Y., Qian, M., Ma, X., Zhu, L., & Martin, J. W. (2018b). Nontarget Mass Spectrometry Reveals New Perfluoroalkyl Substances in Fish from the Yangtze River and Tangxun Lake, China. *Environmental Science & Technology, 52*(10), 5830–5840. https://doi.org/10.1021/acs.est.8b00779

Liu, Z., Lu, Y., Song, X., Jones, K., Sweetman, A. J., Johnson, A. C., Zhang, M., Lu, X., & Su, C. (2019b). Multiple crop bioaccumulation and human exposure of perfluoroalkyl substances around a mega fluorochemical industrial park, China: Implication for planting optimization and food safety. *Environment International, 127*, 671–684.

Macheka, L. R., Olowoyo, J. O., Mugivhisa, L. L., & Abafe, O. A. (2021). Determination and assessment of human dietary intake of per and polyfluoroalkyl substances in retail dairy milk and infant formula from South Africa. *Science of The Total Environment, 755*, 142697.

McCarthy, C., Kappleman, W., & DiGuiseppi, W. (2017). Ecological considerations of per- and polyfluoroalkyl substances (PFAS). *Current Pollution Reports, 3*, 289–301.

Meng, P., DeStefano, N. J., & Knappe, D. R. U. (2022). Extraction and matrix cleanup method for analyzing novel per-and polyfluoroalkyl ether acids and other per-and polyfluoroalkyl substances in fruits and vegetables. *Journal of Agricultural and Food Chemistry, 70*(16), 4792–4804.

Menger, R. F., Funk, E., Henry, C. S., & Borch, T. (2021). Sensors for detecting per-and polyfluoroalkyl substances (PFAS): A critical review of development challenges, current sensors, and commercialization obstacles. *Chemical Engineering Journal, 417*, 129133.

Militao, I. M., Roddick, F. A., Bergamasco, R., & Fan, L. (2021). Removing PFAS from aquatic systems using natural and renewable material-based adsorbents: A review. *Journal of Environmental Chemical Engineering, 9*(4), 105271. https://doi.org/https://doi.org/10.1016/j.jece.2021.105271

Miralles, P., Beser, M. I., Sanchís, Y., Yusà, V., & Coscollà, C. (2023). Determination of 21 per- and poly-fluoroalkyl substances in paper- and cardboard-based food contact materials by ultra-high-performance liquid chromatography coupled to high-resolution mass spectrometry. *Analytical Methods, 15*(12), 1559–1568. https://doi.org/10.1039/D3AY00083D

Miranda, D. de A., Peaslee, G. F., Zachritz, A. M., & Lamberti, G. A. (2022). A worldwide evaluation of trophic magnification of per- and polyfluoroalkyl substances in aquatic ecosystems. *Integrated Environmental Assessment and Management, 18*(6), 1500–1512.

Mullin, L., Katz, D. R., Riddell, N., Plumb, R., Burgess, J. A., Yeung, L. W. Y., & Jogsten, I. E. (2019). Analysis of hexafluoropropylene oxide-dimer acid (HFPO-DA) by liquid chromatography-mass spectrometry (LC-MS): Review of current approaches and environmental levels. *TrAC Trends in Analytical Chemistry, 118*, 828–839.

Munoz, G., Ray, P., Mejia-Avendaño, S., Duy, S. V., Do, D. T., Liu, J., & Sauvé, S. (2018). Optimization of extraction methods for comprehensive profiling of perfluoroalkyl and polyfluoroalkyl substances in firefighting foam impacted soils. *Analytica Chimica Acta*, *1034*, 74–84.

Nakayama, S. F., Yoshikane, M., Onoda, Y., Nishihama, Y., Iwai-Shimada, M., Takagi, M., Kobayashi, Y., & Isobe, T. (2019). Worldwide trends in tracing poly-and perfluoroalkyl substances (PFAS) in the environment. *TrAC Trends in Analytical Chemistry*, *121*, 115410.

Nguyen, H. T., Kaserzon, S. L., Thai, P. K., Vijayasarathy, S., Bräunig, J., Crosbie, N. D., Bignert, A., & Mueller, J. F. (2019). Temporal trends of per-and polyfluoroalkyl substances (PFAS) in the influent of two of the largest wastewater treatment plants in Australia. *Emerging Contaminants*, *5*, 211–218.

Ogunbiyi, O. D., Ajiboye, T. O., Omotola, E. O., Oladoye, P. O., Olanrewaju, C. A., & Quinete, N. (2023). Analytical approaches for screening of per-and poly fluoroalkyl substances in food items: A review of recent advances and improvements. *Environmental Pollution*, *329*, 121705.

Olomukoro, A. A., Emmons, R. V, Godage, N. H., Cudjoe, E., & Gionfriddo, E. (2021). Ion exchange solid phase microextraction coupled to liquid chromatography/laminar flow tandem mass spectrometry for the determination of perfluoroalkyl substances in water samples. *Journal of Chromatography A*, *1651*, 462335.

Pan, Y., Wang, J., Yeung, L. W. Y., Wei, S., & Dai, J. (2020). Analysis of emerging per-and polyfluoroalkyl substances: Progress and current issues. *TrAC Trends in Analytical Chemistry*, *124*, 115481.

Pasecnaja, E., Bartkevics, V., & Zacs, D. (2022). Occurrence of selected per-and polyfluorinated alkyl substances (PFASs) in food available on the European market–A review on levels and human exposure assessment. *Chemosphere*, *287*, 132378.

Perovani, I. S., Barbetta, M. F. S., Duarte, L. O., & de Oliveira, A. R. M. (2023). Determination of polyfluoroalkyl substances in biological matrices by chromatography techniques: A review focused on the sample preparation techniques-review. *Journal of Chromatography Open*, *3*, 100082.

Podder, A., Sadmani, A. H. M. A., Reinhart, D., Chang, N.-B., & Goel, R. (2021). Per and poly-fluoroalkyl substances (PFAS) as a contaminant of emerging concern in surface water: A transboundary review of their occurrences and toxicity effects. *Journal of Hazardous Materials*, *419*, 126361.

Ramírez Carnero, A., Lestido-Cardama, A., Vazquez Loureiro, P., Barbosa-Pereira, L., Rodríguez Bernaldo de Quirós, A., & Sendón, R. (2021). Presence of perfluoroalkyl and polyfluoroalkyl substances (PFAS) in food contact materials (FCM) and its migration to food. *Foods*, *10*(7), 1443.

Rawn, D. F. K., Ménard, C., & Feng, S. Y. (2022). Method development and evaluation for the determination of perfluoroalkyl and polyfluoroalkyl substances in multiple food matrices. *Food Additives & Contaminants: Part A*, *39*(4), 752–776.

Redmon, J. H., DeLuca, N. M., Thorp, E., Liyanapatirana, C., Allen, L., Andrew J., & Kondash, A. J. (2025). Hold my beer: The linkage between municipal water and brewing location on PFAS in popular beverages. *Environmental Science & Technology*, *59*(17), 8368–8379. https://doi.org/10.1021/acs.est.4c11265

Rehman, A. U., Crimi, M., & Andreescu, S. (2023). Current and emerging analytical techniques for the determination of PFAS in environmental samples. *Trends in Environmental Analytical Chemistry*, *37*, e00198.

Ren, J., Lu, Y., Han, Y., Qiao, F., & Yan, H. (2023). Novel molecularly imprinted phenolic resin–dispersive filter extraction for rapid determination of perfluorooctanoic acid and perfluorooctane sulfonate in milk. *Food Chemistry*, *400*, 134062.

Rodriguez, K. L., Hwang, J.-H., Esfahani, A. R., Sadmani, A. H. M. A., & Lee, W. H. (2020). Recent developments of PFAS-detecting sensors and future direction: A review. *Micromachines, 11*(7), 667.

Salihović, S., Dickens, A. M., Schoultz, I., Fart, F., Sinisalu, L., Lindeman, T., Halfvarson, J., Orešič, M., & Hyötyläinen, T. (2020). Simultaneous determination of perfluoroalkyl substances and bile acids in human serum using ultra-high-performance liquid chromatography-tandem mass spectrometry. *Analytical and Bioanalytical Chemistry, 412*, 2251–2259.

Scher, D. P., Kelly, J. E., Huset, C. A., Barry, K. M., Hoffbeck, R. W., Yingling, V. L., & Messing, R. B. (2018). Occurrence of perfluoroalkyl substances (PFAS) in garden produce at homes with a history of PFAS-contaminated drinking water. *Chemosphere, 196*, 548–555.

Schütz, A., Brandt, S., Liedtke, S., Foest, D., Marggraf, U., & Franzke, J. (2015). Dielectric barrier discharge ionization of perfluorinated compounds. *Analytical Chemistry, 87*(22), 11415–11419.

Seo, S.-H., Son, M.-H., Shin, E.-S., Choi, S.-D., & Chang, Y.-S. (2019). Matrix-specific distribution and compositional profiles of perfluoroalkyl substances (PFASs) in multimedia environments. *Journal of Hazardous Materials, 364*, 19–27.

Shahabi Nejad, M., Soltani Nejad, H., Arabnejad, S., & Sheibani, H. (2021). Enhanced adsorption of perfluorooctanoic acid using functionalized imidazolium iodide ionic liquid-based poly (glycidyl methacrylate). *Journal of Applied Polymer Science, 138*(38), 50962.

Song, X., Vestergren, R., Shi, Y., Huang, J., & Cai, Y. (2018). Emissions, transport, and fate of emerging per-and polyfluoroalkyl substances from one of the major fluoropolymer manufacturing facilities in China. *Environmental Science & Technology, 52*(17), 9694–9703.

Song, X.-L., Lv, H., Liao, K.-C., Wang, D.-D., Li, G.-M., Wu, Y.-Y., Chen, Q.-Y., & Chen, Y. (2023). Application of magnetic carbon nanotube composite nanospheres in magnetic solid-phase extraction of trace perfluoroalkyl substances from environmental water samples. *Talanta, 253*, 123930.

Srivastava, P., Williams, M., Du, J., Navarro, D., Kookana, R., Douglas, G., Bastow, T., Davis, G., & Kirby, J. K. (2022). Method for extraction and analysis of per-and poly-fluoroalkyl substances in contaminated asphalt. *Analytical Methods, 14*(17), 1678–1689.

Sunderland, E. M., Hu, X. C., Dassuncao, C., Tokranov, A. K., Wagner, C. C., & Allen, J. G. (2019). A review of the pathways of human exposure to poly-and perfluoroalkyl substances (PFASs) and present understanding of health effects. *Journal of Exposure Science & Environmental Epidemiology, 29*(2), 131–147.

Sznajder-Katarzyńska, K., Surma, M., Wiczkowski, W., & Cieślik, E. (2019). The perfluoroalkyl substance (PFAS) contamination level in milk and milk products in Poland. *International Dairy Journal, 96*, 73–84.

Timshina, A., Aristizabal-Henao, J. J., Da Silva, B. F., & Bowden, J. A. (2021). The last straw: Characterization of per- and polyfluoroalkyl substances in commercially-available plant-based drinking straws. *Chemosphere, 277*, 130238. https://doi.org/https://doi.org/10.1016/j.chemosphere.2021.130238

Vavrouš, A., Ševčík, V., Dvořáková, M., Čabala, R., Moulisová, A., & Vrbík, K. (2019). Easy and inexpensive method for multiclass analysis of 41 food contact related contaminants in fatty food by liquid chromatography–Tandem mass spectrometry. *Journal of Agricultural and Food Chemistry, 67*(39), 10968–10976. https://doi.org/10.1021/acs.jafc.9b02544

Vogel, P., Marggraf, U., Brandt, S., García-Reyes, J. F., & Franzke, J. (2019). Analyte-tailored controlled atmosphere improves dielectric barrier discharge ionization mass spectrometry performance. *Analytical Chemistry, 91*(5), 3733–3739.

Vughs, D., Baken, K. A., Dingemans, M. M. L., & De Voogt, P. (2019). The determination of two emerging perfluoroalkyl substances and related halogenated sulfonic acids and their significance for the drinking water supply chain. *Environmental Science: Processes & Impacts, 21*(11), 1899–1907.

Wang, X. F., Wang, Q., Li, Z. G., Huang, K., Li, L. D., & Zhao, D. H. (2018b). Determination of 23 perfluorinated alkylated substances in water and suspended particles by ultra-performance liquid chromatography/tandem mass spectrometry. *Journal of Environmental Science and Health, Part A, 53*(14), 1277–1283.

Wang, Y., Kim, J., Huang, C.-H., Hawkins, G. L., Li, K., Chen, Y., & Huang, Q. (2022). Occurrence of per-and polyfluoroalkyl substances in water: A review. *Environmental Science: Water Research & Technology, 8*(6), 1136–1151.

Wang, Y., Shi, Y., Vestergren, R., Zhou, Z., Liang, Y., & Cai, Y. (2018a). Using hair, nail and urine samples for human exposure assessment of legacy and emerging per- and polyfluoroalkyl substances. *Science of The Total Environment, 636*, 383–391. https://doi.org/https://doi.org/10.1016/j.scitotenv.2018.04.279

Wang, Y., Yu, N., Zhu, X., Guo, H., Jiang, J., Wang, X., Shi, W., Wu, J., Yu, H., & Wei, S. (2018c). Suspect and nontarget screening of per-and polyfluoroalkyl substances in wastewater from a fluorochemical manufacturing park. *Environmental Science & Technology, 52*(19), 11007–11016.

Wang, Y., Darling, S. B., & Chen, J. (2021). Selectivity of per-and polyfluoroalkyl substance sensors and sorbents in water. *ACS Applied Materials & Interfaces, 13*(51), 60789–60814.

Winchell, L. J., Wells, M. J. M., Ross, J. J., Fonoll, X., Norton Jr, J. W., Kuplicki, S., Khan, M., & Bell, K. Y. (2021). Analyses of per-and polyfluoroalkyl substances (PFAS) through the urban water cycle: Toward achieving an integrated analytical workflow across aqueous, solid, and gaseous matrices in water and wastewater treatment. *Science of The Total Environment, 774*, 145257.

Woudneh, M. B., Chandramouli, B., Hamilton, C., & Grace, R. (2019). Effect of sample storage on the quantitative determination of 29 PFAS: Observation of analyte interconversions during storage. *Environmental Science & Technology, 53*(21), 12576–12585.

Xian, Y., Liang, M., Wu, Y., Wang, B., Hou, X., Dong, H., & Wang, L. (2020). Fluorine and nitrogen functionalized magnetic graphene as a novel adsorbent for extraction of perfluoroalkyl and polyfluoroalkyl substances from water and functional beverages followed by HPLC-Orbitrap HRMS determination. *Science of the Total Environment, 723*, 138103.

Xu, B., Liu, S., Zhou, J. L., Zheng, C., Weifeng, J., Chen, B., Zhang, T., & Qiu, W. (2021). PFAS and their substitutes in groundwater: Occurrence, transformation and remediation. *Journal of Hazardous Materials, 412*, 125159.

Yu, N., Guo, H., Yang, J., Jin, L., Wang, X., Shi, W., Zhang, X., Yu, H., & Wei, S. (2018). Non-target and suspect screening of per- and polyfluoroalkyl substances in airborne particulate matter in China. *Environmental Science & Technology, 52*(15), 8205–8214. https://doi.org/10.1021/acs.est.8b02492

Zabaleta, I., Blanco-Zubiaguirre, L., Baharli, E. N., Olivares, M., Prieto, A., Zuloaga, O., & Elizalde, M. P. (2020). Occurrence of per- and polyfluorinated compounds in paper and board packaging materials and migration to food simulants and foodstuffs. *Food Chemistry, 321*, 126746. https://doi.org/https://doi.org/10.1016/j.foodchem.2020.126746

Zarębska, M., & Bajkacz, S. (2023). Poly–and perfluoroalkyl substances (PFAS)-recent advances in the aquatic environment analysis. *TrAC Trends in Analytical Chemistry, 163*, 117062.

Zenobio, J. E., Salawu, O. A., Han, Z., & Adeleye, A. S. (2022). Adsorption of per-and polyfluoroalkyl substances (PFAS) to containers. *Journal of Hazardous Materials Advances, 7*, 100130.

Zhi, Y., & Liu, J. (2018). Sorption and desorption of anionic, cationic and zwitterionic poly-fluoroalkyl substances by soil organic matter and pyrogenic carbonaceous materials. *Chemical Engineering Journal, 346*, 682–691.

Zhou, Y., Lian, Y., Sun, X., Fu, L., Duan, S., Shang, C., Jia, X., Wu, Y., & Wang, M. (2019). Determination of 20 perfluoroalkyl substances in greenhouse vegetables with a modi-fied one-step pretreatment approach coupled with ultra performance liquid chromatog-raphy tandem mass spectrometry (UPLC-MS-MS). *Chemosphere, 227*, 470–479.

Zou, D., Li, P., Yang, C., Han, D., & Yan, H. (2022). Rapid determination of perfluorinated compounds in pork samples using a molecularly imprinted phenolic resin adsorbent in dispersive solid phase extraction-liquid chromatography tandem mass spectrometry. *Analytica Chimica Acta, 1226*, 340271.

6 Remediation, Treatment, and Management Strategies for Perfluoroalkyl and Polyfluoroalkyl Substances

6.1 INTRODUCTION

As reported by Jin & Zhang (2015), since fluorine has a high electronegativity and C–F interactions are strong, degrading perfluoroalkyl and polyfluoroalkyl substances (PFASs) with traditional biological and chemical methods is somehow difficult. Since PFASs have poor volatility and are resistant to biodegradation, many traditional remediation strategies employed in treating organic composites, such as hydrocarbons and chlorinated solvents are unusable compounds (Glüge et al., 2020; Dawson et al., 2023; Redmon et al., 2025). It has been demonstrated by McGuire et al. (2014); and Dauchy et al. (2017) that certain types of chemical oxidation, air sprinkling, and oxygenation in producing aerobic settings can convert polyfluorinated precursors of perfluoroalkyl (PFA) into PFA compound. The mass flow of PFASs out of treatment regions is projected to rise with the application of these technologies.

The EPA issued the Lifetime Health Advisories of 70 ng/L for the combined concentrations (CCNs) of PFOA and PFOS in drinking water in 2016 (EPA, 2016) in response to the serious public health concern. Most states have implemented or are getting ready to comply with the EPA's health advisory levels for PFOA and PFOS (Pontius, 2019). Nonetheless, some jurisdictions are approaching addressing the issue of PFOA and PFOS in water for consumption from different perspectives. The California State Water Resources Control Board (CSWRCB) declared that its health-based notification levels for PFOA and PFOS were 14 ng/L and 13 ng/L, respectively (CSWRCB, 2019). The Minnesota Department of Health (MDH) chose

DOI: 10.1201/9781003625537-6

somewhat lower health advisory standards for PFOA and PFOS (35 ng/L and 27 ng/L, respectively) than the EPA's health advisory levels (MDH, 2019). A very low health warning limit of 20 ng/L for the combined amounts of PFOA, PFOS, PFHxS, PFHpA, and PFNA was adopted in some severe circumstances, including Vermont (VDH, 2018). It should be mentioned that in order to develop a new water consumption regulation, the expenses of achieving the maximum contamination level (MCL) must be justified using the water treatment and management technology now in use. Unsurprisingly, it is very difficult to degrade PFOA and PFOS from the environment to below the aforementioned advisory levels using conventional methods of remediation, treatment, or management like soil vapour extraction, air stripping, or chemical oxidation because of the persistent chemical stabilities and unique molecular structures (Kucharzyk et al., 2017; Lu et al., 2020). As a result, one of the most prevalent topics in the science and technology of the environment is advancing the development of suitable degrading techniques to economically accomplish the objective of PFAS management and remediation. Adsorption (AST) (Lin et al., 2015; Milinovic et al., 2015), filtration (Tang et al., 2007), sonochemical destruction (Fernandez et al., 2016; Lu et al., 2020), electrochemical oxidation or electrochemical anodic oxidation (EO) (Niu et al., 2016; Trautmann et al., 2015; Zhang et al., 2016; Zhuo et al., 2020), photocatalytic oxidation (Gomez-Ruiz et al., 2018 ; Sahu et al., 2018), electron beam (EB) destruction (Kim et al., 2019), persulfate oxidation (Zhang et al., 2019), plasma-based treatment (Jovicic et al., 2018; Mahyar et al., 2019; Singh et al., 2019), and biological remediation, treatment, or management (Dinglasan et al., 2004; Natarajan et al., 2005; Lu et al., 2020), among other advanced remedial, treatment, and management measures based on various technologies, have been reported in numerous studies.

The majority of these emerging innovations and technologies necessitate severe conditions for operation, centralized laboratory apparatus, and substantial chemical and energy consumption, all of which raise the expense and complexity of the treatment and management procedure and diminish its application and practicability. A current trend in academia is connecting or combining several remedial and management treatments as an approach train that will enhance the entire efficacy and practical possible outcomes, in addition to the growing public awareness and emerging remediation technology, see Lu et al. (2020) for more details. A summary of published treatment and management train studies on PFAS remediation and management is provided in this chapter, along with information on their creative concepts, remediation outcomes, current constraints, and future directions. According to Lu et al. (2020), most of all the chosen treatment and management trains fall into one of two groups: (a) tandem arrangement of a degradation technology after a removal approach, or (b) parallel arrangement of several contemporary degradation techniques, as shown in Figure 6.1.

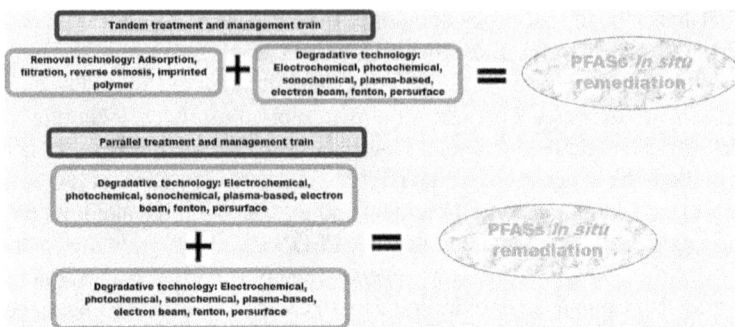

FIGURE 6.1 Diagrammatic representation of two main treatment and management train approaches for PFAS remediation and management.

6.2 BIOLOGICAL TREATMENT, REMEDIATION, OR MANAGEMENT APPROACHES OF PFAS

Although the rate of this transition may be sluggish and some of the temporary intermediates produced are not yet characterized, polyfluorinated categories of PFASs could possibly biotransform in the ecosystem to produce PFA (Benskin et al., 2013; Dasu et al., 2013; Lee et al., 2014; D'Agostino & Mabury, 2017).

Typically, bacteria need a minimum of a single hydrogen atom at the α-carbon next to the perfluoroalkyl (PFA) chain for the first assault in order to proceed with the biotransformation of PFASs (Key et al., 1998). Several polyfluorinated precursor compounds of PFA biotransform into chain PFA that are shorter (Liu & Avendaño, 2013; Harding-Marjanovic et al., 2015) when compared to the original polyfluorinated compounds' PFA chain length. For instance, 8:2 fluorotelomer alcohol (FTOH) undergoes aerobic transformation to produce PFA compounds, such as "perfluorooctanoic acid (PFOA) and perfluorohexanoic acid (PFHxA)" (Wang et al., 2005). When the composites of fluorotelomer (FT) biotransform to PFA compound, there is a partial shortening of the PFA chain; nonetheless, no mineralization is noted, and stoichiometric amounts of PFA compound would be generated from the FT precursors.

Also, *Phanerochaete chrysosporium*, which is a white-rot fungus, was shown to biotransform 6:2 FTOH during the course of 28 days of incubation (Tseng et al., 2014); perfluoropentanoic acid (PFPeA), PFHxA, and a combination of 5:3 FT acid, were found to be the main transition products. Fungal biotransformation of PFOA has been documented (Luo et al., 2015; Colosi et al., 2009). The content of PFOA was reduced by 30% when the horseradish peroxidase enzyme was used as a fungal enzyme treatment. As reported by Colosi et al. (2009), a reactive phenolic cosubstrate was oxidized by the enzyme horseradish peroxidase, which resulted in phenolic radical interactions with PFOA that produced shorter-chain molecules. It has been reported that in 157 days, PFOA was broken down into partly fluorinated-chain alcohols that were shorter and the aldehydes were also broken down by another enzyme-catalyzed oxidation process utilizing laccase as the enzyme (Luo et al., 2015); conversely, it is uncertain what these suggested breakdown intermediates are.

At the moment, there are hardly any reports of fungi attacking PFSAs like PFOS by enzymatic means. From some of the reported studies, PFASs have not yet been shown to biodegrade (Ochoa-Herrera et al., 2016; Luo et al., 2015; Liu & Avendaño, 2013).

According to the Organisation for Economic Cooperation and Development (OECD), the process through which organic materials are broken down into elements like carbon dioxide, water, and ammonia by microorganisms (mostly aerobic bacteria) is known as biodegradation (OECD, 2008). Since there is no evidence that PFASs mineralize—a process that produces stoichiometric amounts of fluoride, carbon dioxide, and maybe sulfate for the PFA sulfonates as well as their precursors—PFASs reportedly do not biodegrade. Despite certain instances of biological or organic attacks on PFASs being reported as biodegradation, none of the roughly 3,000 PFASs can biodegrade (Dasu et al., 2012, 2013, 2016; D'Agostino & Mabury, 2017). The precursors of PFA biotransform to produce PFA compounds that are persistent.

The employment of enzymes in soil (terrestrial) or water (aquatic) management and treatment raises several questions for commercial applications. PFA sulfonates, like PFOS, have not been shown to be treated. It has been claimed that the fungal laccase enzyme may possibly convert 50% of PFOA to PFPeA in 157 days (a stoichiometric release of fluoride of 28% was detected) (Luo et al., 2015). Enzymic activity occurs at a pace that is considerably too sluggish for commercial use; in ideal laboratory settings, about 50% of PFOA biotransformation takes over five months. Since mineralization has not been proven, this treatment method is believed to be unrealistic when taking into account the retention times of hydraulics in wastewater management and treatment systems or commercial durations for soil management and treatment. Particularly, this is true when additional mobile, shorter-chain PFA compound would possibly result from an enzymatic attack. Given that enzymes are basically proteins as well as a very rich source of nutrition for some other bacteria, there may be concerns about the stability of enzymes in commercial management and treatment settings. Recently, attempts were made to safeguard the enzyme in subcellular organelles known as vaults; nonetheless, the biotransformation of PFOA proved unsuccessful (Mahendra et al., 2016), and there would probably be a large cost associated with using vaults. For the laccase enzymes to continue biotransforming PFOA, they also require an oxygen and cofactor supply, which supplicates the question of how they will be obtained in practical applications. Evaluation of the effectiveness of white-rot fungus and the general excretion of fungal enzymes for the management and control of pollution has advanced (Gao et al., 2010, 2018, 2020). However, the numerous PFAS-specific difficulties previously mentioned show that considerable potential obstacles must be surmounted in terms of cost, schedule, and practicality before thinking of fungal management and treatment methods as suitable for large-scale commercial PFAS cleanup procedures.

An existing wetland was used to test for the phytoremediation of certain PFASs, however, the results indicated no discernible elimination of PFAS compound (Plumlee et al., 2008, 2009). Evaluation is needed in view of concerns raised about the ecotoxicological impacts of PFA leaks into some delicate regions of the ecosystems. Some research on edible crops and plants as well as soil sorption does show that there are active processes for absorption and sorption; long chains tend to concentrate in the roots as well as the shoots, whereas short chains tend to concentrate in fruits (Blaine et al., 2013, 2014a, b). Using a range of tree species and native plants, such as "silver birch, Norway spruce, bird cherry, mountain ash, ground elder, long beech fern, and wild strawberry", research was conducted on numerous plants' capacity to concentrate 26 PFASs (Gobelius et al., 2017). For PFOS (beech fern), the highest bioconcentration (bio-CCN) factors were recorded as 906, and for PFOA (spruce), as 41. According to this study, it is unlikely that PFAS bio-CCN in tree species would reach metal quantities, where some hyperaccumulators of nickel, for instance, may collect 26% nickel on a basis of dry weight (Jaffré et al., 1979). According to reports from Gobelius et al. (2017), the combined 26 PFASs in a tree might weigh up to 11.00 mg for birch and 1.80 mg for spruce. In comparison to metals, the amount of PFAS collected from each tree seems to be relatively little. It could be possible to employ phytoremediation to stop PFASs from entering groundwater plumes, but first groundwater flow has to be studied to see if phytoextraction rates are high enough to control PFAS mass flux in the aquifer.

Since PFASs' PFA chain is chemically reduced and strongly halogenated, they will act as the terminal electron acceptor (TEA) in metabolic procedures from a thermodynamic standpoint. The microbial metabolism of PFA compounds may have some resemblance to the well-established biological reductive dechlorination (DCN) pathways for organochlorine (CNC) compounds, including perchloroethene. When an electron source is available, the anaerobic biodegradation of PFASs as TEAs might seem plausible from a thermodynamic point of view. Moreover, it has been noted that mineralization is not impeded by the thermodynamics of some reported PFASs' biodegradation (like hexafluorethane or octafluororopane) (Parsons et al., 2008). Notably, while being one of the strongest organic bonds and having dissociation energy of about 5.36×10^5 J/mol, the C–F bond is, nonetheless, vulnerable to enzyme assault, as demonstrated by the case of fluoroacetate (Goldman, 1965). Nevertheless, there is no evidence of a biological attack on trifluoroacetate (Matheson et al., 1996), proving that the presence of many fluorine atoms on even the most basic particles of PFA makes them difficult for microbes to assault. As more atoms of fluorine attach to carbon, the intensity of the C–F bond is said to grow; for example, the heat required to form the C–F bond is said to rise from 4.48×10^5 J/mol for CH3F to 4.86×10^5 J/mol for CF4 (Kissa, 2001). Furthermore, it has been shown that a C–F bond's extremely small atomic radius allows it to buffer a perfluorinated carbon atom from steric stress. The three firmly-bound lone pairs of electrons per atom of fluorine, as well as the negative partial charge act as an active electrostatic as well as a steric shield against any form of nucleophilic attack that is directed against the carbon atom that is in the centre, leading to the description of the shielding as a "coating" of the substituents of fluorine giving kinetic stability (Kirsch, 2004). The number of substituents of the fluorine attached to the same atom of carbon was also shown to boost the extreme stability of fluoroorganic compounds. The dimension of the C–F bond has been estimated to decrease from 1.40×10^{-14} picometres (pm) for CH3F to 1.33×10^{-14} pm for CF4, which is said to represent this increase

(Kirsch, 2004). According to reports, this permits almost ideal overlap between the carbon corresponding orbitals and the fluorine 2s and 2p orbitals. This allows for the formation of a dipolar resonance structure for multiple fluorine-substituted carbons, thereby facilitating the "self-stabilization" of several substituents of fluorine on a single atom of carbon.

Other notable distinctions between the compounds of organofluorine (ONF) and CNC include their comparative natural abundance. The intensity of the C–F bond and the stability of numerous C–F bonds in PFA compounds are crucial factors regarding their persistence. Over geological time, CNC chemicals have now been found in the environment from both sources of biogenic as well as volcanic exudates (van Pée & Unversucht, 2003). Over 2,000 CNC chemicals can be found in nature (Gribble, 2002). Hence, there is now enough time for exposure for microbes to develop the ability to digest them. Compared to CNC compounds, the variety of the natural ONF compounds that are now occurring appears to be much smaller, despite the fact that a small number of ONF chemicals have also been documented to be extruded from volcanic activity (Gribble, 2002; Harper et al., 2003); reportedly, there are just around 30 known instances of naturally occurring ONF compounds, of which none are perfluorinated. The larger redox potential needed to produce F+ from F– compared to that needed in forming other ions of halonium from their respective halides has been mostly attributed to the fluoride ion's higher heat of hydration in comparison to other halide ions. This distinction has been explained as preventing the haloperoxidase process, which has been identified as a primary pathway for the formation of organohalogens in nature, from incorporating fluorine into natural compounds (Harper et al., 2003). Since there are hardly any analogues that occur naturally in promoting the production of the necessary catabolic enzymes, microbes have not been opportune to advance the digestion of PFASs by building enzymatic arrangements in attacking them. Furthermore, methanogenesis may take place in preference to defluorination (DFN) in circumstances where reductive DCN could possibly be feasible in a highly reducing aquifer setting, like those that could be necessary to permit reductive DFN.

There would possibly be some intermediate successive DFN in breakdown products if the reductive DFN procedure possibly will be "force evolved", employing uncovered microbial communities by applying a selection pressure (e.g., giving a PFA as a single TEA). During the successive removal of 17 fluorine atoms, microbes would require to become accustomed in order to defluorinate not just PFOS or PFOA, perhaps, but also the intermediate breakdown products. The sluggish development typically observed via the metabolism of halogenated composites and the requirement to cleave a considerable amount of fluorine particles from the various intermediate breakdown products formed make the development of a metabolic route to mineralize PFA compound under anaerobic circumstances seem implausible.

Furthermore, it appears that anaerobic biodegradation of PFASs has limited potential because there are hardly any organisms that have been found to employ PFA compounds as TEAs. Additionally, there could possibly be some obstacles in the way of this process being a practical remediation technique.

6.3 WATER TREATMENT, REMEDIATION, OR MANAGEMENT TECHNOLOGIES OF PFAS

A scatter plot (Figure 6.2) linking the phase of research as well as the range of practicality has been used to establish accessible remedial or treatment options for water that are pertinent to PFASs.

Remedial/treatment technologies' efficacy against several PFAS groups is examined rather than concentrating just on two or three PFAAs. A summary of many technologies that are being used, expanding in scope, or recently coming out of lab testing is provided by these authors (Lu et al., 2020; Garg et al., 2023). A review of frequently used interim corrective actions, such as granular activated carbon (GAC) and ion-exchange resins (IXs), is part of this advancements. Precipitation techniques exist to reduce elevated levels of PFASs before employing adsorbents, hence, extending their useful lives. However, these techniques also result in precipitant waste sludge and dewatering difficulties. Academic creation of innovative technology or adaptation from usage for other pollutants are the two ways in which new adsorbents are developing. A critical assessment of the prospective applications of technologies recently utilized for PFAS separation, such as foam fractionation (FF), is given. Since these techniques are being promoted as in situ PFAS remedies, the use of FF in situ as well as the inoculation of particulate-activated carbon (AC) are also examined (Lu et al., 2020; Garg et al., 2023). Nevertheless, all these techniques do is transfer PFASs from one matrix to another, not eliminate them. Consequently, some thought is given to technologies that may be paired with other technologies in a treatment train and are purportedly capable of destroying PFAS. The goal is to present a technical study that addresses the viability of deploying remedial solutions while taking into account geological and hydrogeological aspects as well as the chemistry of PFASs.

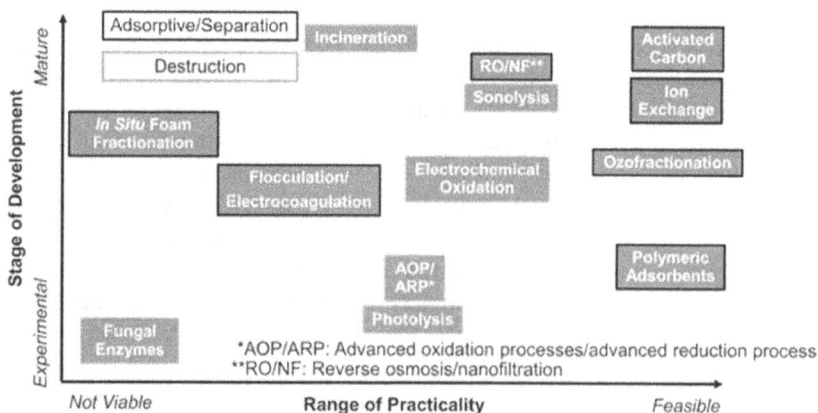

FIGURE 6.2 A scatter plot of the remedial, treatment, or management technologies for PFAS in water.

Due to the possibility for reverse diffusion of toxins from less conductive aquifer horizons, the utilization of groundwater pumping for cleanup has been reported to be limitless. The comparatively high solubility of certain PFAAs, the occurrence of huge diffuse plumes, and the extremely low precipitation (ppt) treatment objectives that may be applied to PFASs raise concerns about the treatment and remedial performance on a long-term basis when pumping groundwater to confiscate PFASs. It appears that groundwater pumping may be used for the management and control of PFASs for many years to come. Concerns regarding the feasibility of long-term pump and treatment projects were raised by the quantity of energy and resources needed to manage PFASs over an extended period of time by groundwater pumping.. Groundwater extraction methods that compensate for sluggish back diffusion have seen significant improvements in the last ten years, enabling quicker aquifer cleansing (Lu et al., 2020; Garg et al., 2023; Potter, 2016; Suthersan et al., 2015a, b). The use of precisely planned groundwater recirculation systems, also referred to as dynamic groundwater recirculation (DGR), has made this possible (Suthersan et al., 2015a,b; Lu et al., 2020; Garg et al., 2023). Increasing the pore volume flushing of water that is clean and specifically reinjected after cleansing and treatment, as well as figuring out the natural conveyance mechanisms that cause contaminants to be stored in an aquifer, are two aspects of DGR performance. Pollutant flushing is accelerated by multidirectional groundwater flushing across an affected aquifer using several pore volumes. The injection process is tailored to the unique geology and hydrogeology of the site, directing flow in the direction of extraction while minimizing the amount of unaffected groundwater recovered and optimizing mass recovery. DGR is not the same as in-well groundwater recirculation, which usually has smaller than predicted radii of effect and displays short-circuiting within the well (Allmon et al., 1999).

Adaptation is essential while utilizing DGR in order to maintain a situation of instability between the conveyance and storage regions and quicken the multidirectional flushing in the advective conveyance regions. DGR is a dynamic procedure that encompasses adaptively, altering the reinjection as well as the extraction outlines on a regular basis. This increases the rates of back-diffusion of stored pollutant mass through an improved advective flux formed by reinjecting treated or purified water. Large plumes where traditional treatment methods would be impractical or prohibitively expensive to implement can be remedied using DGR because PFAS plumes frequently demonstrate distortion between the centre of mass and the discharge point. When pumping water from PFAS-based source sites as well as the plumes, the adoption of DGR should significantly reduce treatment timeframes. Concerns about partially treated water being reinjected will be lessened with the implementation of further detailed treatment or remedial methods for the removal of various PFASs from the removed water.

6.3.1 GAC REMEDIATION, TREATMENT, OR MANAGEMENT METHODS OF PFAS

At the moment, GAC is one of the common treatment methods used for water in eliminating PFOS and PFOA, as well as in reducing the degree of contaminations and other PFA composites from water (Lu et al., 2020; Garg et al., 2023; Merino et al., 2016; Du et al., 2014). It is a proven mechanism that could be possibly used as a stand-alone unit or as a component of a treatment train, and it can be implemented at scales ranging from home point of entry systems to municipal water treatment facilities. With an effectiveness rate of about 90%, GAC could reliably confiscate PFOS at quantities of portions per billion or micrograms per litre (µg/L) (Eschauzier et al., 2012; Oliaei et al., 2013). PFOA can be somewhat difficult to remove using GAC, though (Oliaei et al., 2013; Lu et al., 2020; Garg et al., 2023). Consequently, as the chain length decreases, it becomes less and less efficient at eliminating shorter-chain PFCAs such as "PFHxA, PFPeA, PFBS, and PFBA" (McCleaf et al., 2017; Inyang & Dickenson, 2017; Lu et al., 2020; Garg et al., 2023). GAC systems are now the standard wherein all other adsorbent procedures aimed at removing PFAS from water are adjudicated.

There have been reports of PFA removal in large-scale water remedial or treatment systems (Lu et al., 2020; Garg et al., 2023; Appleman et al., 2014). According to Appleman et al. (2014), over the course of the last few years, the breakdown of five PFA compounds was seen in a municipal water treatment facility that was built specially to use GAC for the confiscating of small parts per billion PFA CCNs. With Calgon Filtrasorb 600 (F600) GAC and a 13 min (780 seconds) empty bed contact time (EBCT) in a lead-lag arrangement comprising two GAC containers, the system was intended to treat 1.40 to 1.50 m^3/minute. Prior to the discovery of PFA, 6.00×10^4 bed volumes (BVs) were treated for PFOS, 3.00×10^4 BVs for PFHxA and PFOA, and 5.00×10^3 BVs for PFBA.

Eschauzier et al. (2012) observed comparable breakthrough patterns in PFA for lead-lag setup, 348 m^3/hour, 20 minutes (1.20×10^3 seconds) EBCT, influent water with low ppt PFA compound. In treated water, both PFBA and PFHxA eventually demonstrated breakthrough levels above their influent CCNs. The PFOS and PFOA CCNs in the effluent (waste water) from the former lag container decreased significantly, while the PFHxA CCNs decreased by more than fifty per cent. The PFBA CCNs remained unchanged. This was achieved by changing the configuration of the procedure in moving the lag container into the lead location and substituting the carbon in the original lead container. Based on these findings, the authors hypothesized that desorption and release of the earlier adsorbed PFHxA and PFBA were caused by rivalry for sorption locations on the GAC from longer-chain PFASs and/or naturally occurring organic materials. For PFBS, comparable desorption behaviour has been seen (Eschauzier et al., 2012).

In a GAC Calgon Filtrasorb 300 (F300) column investigation with 1 µg/L PFA compound, PFA carboxylates showed advancement faster than PFA sulfonates of similar PFA chain length. PFA sulfonate chain length "(PFOS < 98,000 BVs, PFHxS < 45,000 BVs, and PFBS < 30,000 BVs)" decreased the number of BVs treated before breakthrough (Higgins & Luthy, 2006; Ahrens et al., 2010). Also, the study discovered that the presence of natural organic matter (1.70 mg/L in stream water) strongly reduced the removal efficacy of all PFA compounds, together with the long-chain species like perfluorononanoic acid. Additionally, competition from naturally existing organic matter, the presence of additional organic composites with comparable molecular weights as well as higher sorption potential may hinder PFAS AST. The need to conduct water column research in the field is underscored by the competing AST from other co-occurring composites in the inflowing. Evidently, for reports under our disposal, there are no studies that have been reported on the efficiency of GAC for confiscating anionic, zwitterionic, or cationic precursor compounds. But according to a new theoretical analysis, GAC might not be able to eliminate every precursor (Xiao et al., 2017).

It has been reported by Ochoa-Herrera & Sierra-Alvarez (2008) and Zhi & Liu (2015) that GAC surface area, pore size, and surface chemistry are linked to the AST ability for PFOS and PFOA. According to reports from the studies by Rattanaoudom et al. (2012) , powdered AC (PAC) works better than GAC. However, the assertion is generally and conclusively not acknowledged (Zhi & Liu, 2015; Sun et al., 2016). It is crucial to note that PAC is normally not revived, hence, the manufacturer might possibly not be able to recover expended PAC. It has been demonstrated in the studies of Deng et al. (2013) and Zhi & Liu (2015) that GAC based on wood and bamboo performs better than coal-based composites, with AST declining at both acidic and alkaline pH levels. Over the past few years, several types of coconut shell carbon, such as over-activated coconut carbon, have been developed with the goal of enhancing PFA removal.

According to recent studies, during the reactivation process, certain PFA compounds could possibly be degraded on GAC surfaces at certain temperatures of about 7.00×10^2 °C. While a temperature of 1.10×10^2 °C is necessary for the obliteration of volatilized PFA compound (in the air phase), thermal re-activation kilns typically have afterburners for air contamination management and control, and these typically run at temperatures that are higher than 1.10×10^2 °C. Accordingly, it appears that reenergizing GAC that has outlived its AST ability for PFA compound may be accomplished effectively using a standard thermal reactivation technique, which involves an afterburner and a temperature range of 8.00×10^2 to 1.00×10^3 °C (Watanabe et al., 2016). Testing, however, was not done taking into account the broader range of PFASs, such as the polyfluorinated precursors with a larger molecular weight (less volatile) that have been linked to AFFF formulations (Backe et al., 2013; Choyke et al., 2017). There is hardly enough reported information on whether these temperatures eliminate all PFASs, together with precursors that could have been adsorbed on GAC.

Summarily, GAC could be an appropriate water treatment technique when long-chain PFA removal is necessary. Its capacity to eliminate PFA precursors is mainly unknown, and its removal of short-chain PFA compound is less reliable. Also, if natural organic matter or co-contaminants are present in the waters that need to be purified or treated, using GAC might not be practical. Prior to implementing GAC, preplanned, small-scale, fast-column testing must be taken into account. Due to the necessity of replacing it, continued usage of GAC is anticipated to result in high costs. Given the variety of PFASs, it seems reasonable to do more testing on the effectiveness of regeneration.

6.3.2 Injectable Particulate Carbon

It has been suggested by Regenesis (2017), that aquifers be injected with customized AC products to address various kinds of dissolved phase pollutants. In order to help the contamination adsorb onto the injected adsorbent, this type of "trap-and-treat" method disperses an adsorbent particle material throughout the aquifer. The idea proposes that in the case of pollutants that are biodegradable, the bulk flow of the contaminant is reduced while biological breakdown is promoted. Because there are no known biodegradation pathways for PFASs, the pollutants will saturate the AC that has been put in. Similar to how GAC is less successful in eliminating PFA precursors and shorter-chain PFA compounds from water (Xiao et al., 2017), it is also anticipated that the injected particulate-AC will be less successful in eliminating these PFASs. After the PFASs have saturated the AC, they may create a secondary source zone and release more PFASs. Several particulate-ACs could be injected to boost the capacity of in situ treatment, however, pore space is restricted, and particulate-AC straining could significantly impede homogeneous distribution in situ.

Hydraulic fracturing, uneven particulate distribution in the targeted treatment or remedial region, and the straining of particles larger than a millimetre in pore spaces are obstacles to effectively treating aquifers using particulate injection. The injected AC, which has a particle dimension of around 1 to 2 microns, would, at certain spots, be squeezed out via the aquifer's pore throats, which will negatively and severely affect the distribution that can be achieved. In the changing regulatory environment surrounding PFAS, injected particulate carbon could not offer sufficient treatment since AC is less efficient at eliminating shorter-chain PFA compounds and PFA precursors. When contaminant breakout is noticed, GAC employed in ex situ treatment systems may be quickly replaced. If inadequate pollutant capture is accomplished, the solution can also be simply changed. However, it is difficult to modify or replace injectable particulate-AC with new therapeutic technologies. Furthermore, much more research is required to fully comprehend the distribution and effectiveness of injectable particulate-AC in treating or managing a variety of PFASs.

6.3.3 Ion-Exchange Resins

A variety of IXs with various functional groups (FGs) that allow for various kinds of selectivity were evaluated for the removal of a limited amount of PFASs from water (Du et al., 2014). More innovative resins are believed to have better sorption efficiencies in both the long- and certain short-chain PFASs when compared to that of GAC, despite the fact that some IXs are applicable for either long- or short-chain PFASs (Zaggia et al., 2016). IXs have the capacity for larger AST efficiencies, shorter contact durations, lesser equipment footprints, as well as the capacity for regeneration, making them more advantageous for particular applications even though they are more costly by weight than GAC and sometimes require pretreatment (Dickenson & Higgins, 2016; Merino et al., 2016). PFAS elimination also makes use of one-time IXs that don't need to be renewed. IXs can be used as a polishing phase in the remediation, management, or treatment train arrangement following GAC.

Resins are generally grouped into two categories: IXs and non-IXs:

- IXs are structurally made of synthetic polymeric nature attached to a polystyrene or polyacrylic bead, with a charged FG balanced by a counterion. The primary method of PFAS removal is the exchange of counterions, which results in electrostatic contacts. However, alternative mechanisms of removal, such as hydrophobic interactions resulting from agglomerated PFASs inside the IXs, are also noteworthy (Zaggia et al., 2016).

* The non-IXs are structurally made of neutral synthetic polymeric nature. Since they lack exchangeable ionic sites, they attach substrates via interactions of non-ionic settings such as hydrophobic as well as the van der Waals contacts after the substrate migrates into the matrix under the influence of diffusion. Strongly alkaline environments yield the best performance from non-IXs (Dow, 2016). Compared to IXs, non-IXs often result in poorer substrate adsorptive binding, making resin regeneration easier (Senevirathna et al., 2010).

Most research has generally employed anion-exchange to remove anionic PFA compounds, such as PFOS and PFOA (Zaggia et al., 2016) as well as non-IXs such as DowV493, which in bench-scale experiments demonstrated an order of magnitude with more PFOS AST capability than GAC, as reported by Senevirathna et al. (2010). The equilibrium kinetics of resin were much extended at 80 hours (4.80×10^3 mins) than they were for the GAC at four hours (240 mins), which might result in a considerable reduction in AST capacity than the traditional ones. In comparison to GAC, regeneration could probably be necessary for the cost-effectiveness of resin. Long-chain PFA and precursors are probably confiscated from the water much more strappingly than short-chain PFASs because non-IXs bind by hydrophobic contacts.

For several common pollutants, IXs are also a proven technique in the groundwater and municipal treatment sectors. IXs, which have been modified for the improvement of their polyatomic ion selectivity, are used to remove sulphate, chromate, nitrate, chloride, and perchlorate. There is a lot of rivalry between these naturally occurring co-contaminants and PFASs for AST sites since many of them are found in aquifers or municipal effluent streams at CCNs orders of degrees higher than PFASs. The possibility of large quantities of entire dissolved solids that might have a significant impact on ionic strength and complicate the electrostatic AST of PFASs onto the IXs, is another issue, particularly for groundwater.

IXs can be regenerated using brine solutions or liquid rinses, including methanol. Although considerable regeneration may be accomplished, the existing price structure favours offshore cremation of single-use IXs due to the challenges and risks associated with concentrating a liquid waste stream. Finally, from available publications, there is presently no reported research on the confiscation of cationic and zwitterionic PFASs utilizing the resin approach. These compounds are not likely to be significantly bound by anion-exchange resins, although they could be eliminated via the interactions of hydrophobic and non-IXs. Hence, a resin technique still requires a remediation, management, or treatment train of several resins in the removal, remediation, management, or treatment of various charged species if a water sample has a mixture of anionic, cationic, and zwitterionic PFASs, as could possibly be predicted in certain PFAS fire-training areas (FTAs) groundwater.

6.3.4 OTHER ADSORBENTS

It has also been demonstrated that chars, ash, and carbon nanotubes are other adsorbents that can absorb PFASs (Chen et al., 2011), fibres made of AC (Zhi & Liu, 2015; Zhi & Liu, 2015; 2018), Ambersorb (Zhi & Liu, 2015), hydrotalcite (Rattanaoudom et al., 2012), imprinted polymers (Yu et al., 2008), cross-linked cyclodextrins (Xiao et al., 2017b), modified cotton and rice husk (Deng et al., 2013, 2012), as well as permeable aromatic frameworks (Luo et al., 2015).

Osorb is one of the interesting novel materials that shows promise for PFAS AST and absorption, which is a polymeric structure based on silica that was developed by ABS Materials, composed of alkoxysilicanes that are cross-linked. When the Osorb structure is exposed to organic molecules, it expands between three and five times its initial volume. This results in the organic composites being absorbed into the mostly microporous matrix, as opposed to being adsorbed. As of right now, Osorb may be purchased either pure or coated onto silica under the brand name Purasorb®, which may be more useful for remediation systems handling lower pollutant CCNs. There are presently no reports of these technologies being used for commercial purposes outside of laboratory-scale research. Adsorbents with strong sorptive capacity, regeneration potential, and the capacity to adsorb a wide range of PFASs may be crucial for their successful commercial adoption.

6.3.5 PRECIPITATION AND SEDIMENTATION

Cornelsen Umwelttechnologie GmbH developed a unique precipitation and sedimentation method called PerfluorAd® for PFOA and PFOS. It is intended for liquid waste streams with greater CCNs (> 0.3 µg/L). A coagulant that uses hydrophobic and electrostatic interactions to adsorb PFOA and PFOS is added as part of the procedure. The precipitate is gathered as waste sludge, sieved, as well as disposed of (Cornelsen, 2017). It is the goal of extending the life of the subsequent remediation, treatment, or management (AST polishing) to reduce the CCN of PFOA and PFOS by some orders of degrees before polishing the water. Significant amounts of waste sludge may be produced, which calls for burning, depending on the coagulant CCNs used, which could possibly range from 25.00 mg/L to 2.00 g/L. Dewatering the waste sludge is another possible issue. This can result in inefficient incineration and is a typical challenge for most wastewater treatment facilities (WWTPs). It is noteworthy that applying greater dosages of the coagulant enhances the precipitation clearance rates for shorter-chain PFA compounds. For instance, the observed PFBA clearance rates ranged from 6% to 30% at lower coagulant doses, with a maximum of 77% (2.00 g/L additive).

6.3.6 OZOFRACTIONATION

Using the ozofractionative catalyzed reagent addition (OCRA) technique, ozone (O_3) is produced (Dickson, 2013, 2014). By means of chemically oxidizing the organic pollutants, in addition to producing concentrated foam fractionates that could possibly be isolated from the treated or purified water, liquid waste is treated.

This system comprises of a series of columns where fine O_3 bubbles are introduced into affected water, oxidizing non-fluorinated organic molecules and removing PFASs. Concentrated in the fractionate stream at the upper part of the column, PFAS is retrieved for off-site disposal or further treatment. Small quantities of recoverable PFAS-enriched foam are produced by the OCRA method, which employs O_3 in micron-sized (< 200 micrometres) gas bubbles to selectively partition PFASs into the high surface area of the microbubbles. The microbubbles rise to the upper part of the container (Evocra, 2016). While the O_3 could concurrently oxidize and break down organics like hydrocarbons from petroleum, the PFA groups in PFASs specially move to the interface of the gas–fluid, allowing for efficient confiscation from the aqueous stage by FF.

The method is a multiphase, adaptable procedure that extracts PFASs such as PFOS, PFHxS, PFOA, PFHxA, PFBS, PFPeS, as well as all other chain-length composites, together with the short-chain precursors and other inorganic co-contaminants from obstructed water (sewage wastewater, groundwater), while oxidizing organic co-contaminants concurrently. Through chemical oxidation, the system may transform polyfluorinated precursors into PFA compounds, which help with their confiscation through FF. It can also confiscate the short-chain PFASs together with the long-chain PFASs.

Waste streams that contain up to 20% solids, such as scum and sludge could be fed into the system for management, remediation, or treatment. Liquids and solids are separated using the fractionation columns. The fractionate stream or foam concentrate removes fine particles that rise to the top of the column. Larger, coarser particles settle towards the bottom of the column and separate by sedimentation. To reach the appropriate discharge levels (typically ng/L), water is put via a polishing stage utilizing reverse osmosis (RO), nanofiltration (NF), or AST technologies. More than 99.96% of long-chain PFA chemicals, including PFOS and PFOA, have been confiscated using the OCRA system (Evocra, 2016).

OCRA has been shown to be capable of fully treating wastewater and surface water contaminated with FT foam (Ross et al., 2017). The system's primary parts, which are contained in transportable containers and are found in Australia, Brisbane, include a feed tank, reaction containers, O_3 generator, as well as sand filter. The efficacy of OCRA for the elimination or removal of a whole spectrum of PFASs, together with PFA compound and PFA precursor chemicals, was evaluated using the TOP test. With a range of input values from 1.00×10^2 µg/L to 5.40×10^3 µg/L total PFASs, more than 97% confiscation or removal of the sum of 28 PFASs determined post-TOP Assay was shown (Ross et al., 2017).

The ability of the OCRA process to break down and confiscate hydrocarbons from petroleum as well as other co-contaminants, like sewage liquid with high BOD and TOC, without sacrificing PFAS removal efficiency, is a key advantage over AST approaches. OCRA has demonstrated that it is extremely configurable and capable of handling several PFAS CCNs. It also enables complicated, multicontaminant confiscation in a very compact footprint. The system in Brisbane is a competent CCN/separation procedure that operates independently of influent CCN, with influent levels ranging from less than 1 µg/L to more than 5000 µg/L.

As previously mentioned, the OCRA process offers a lot of benefits, but it also produces a concentrated effluent stream that has to be managed—the foam concentrate. Usually ranging between 0.50% to 2% of the influent volume, the fractionate or concentrated PFAS waste stream contains PFASs at CCNs more than 1,000 times. In order to polish as well as eliminate PFASs, the OCRA system alone may also find it difficult to attain the extremely low (ng/L) regulation CCNs. As a result, there is a need for modification so as to function as a remediation, management, or treatment train procedure in conjunction with multiple additional technologies. As a component of a remediation, management, or treatment train, the OCRA method concentrates PFASs while maintaining their aqueous phase, which has certain benefits for further harmful treatment. The OCRA process needs more frequent supervision than more mechanically straightforward remediation, management, or treatment technologies, such as the usage of GACs or IXs, however, the utilization of telemetry in monitoring the procedure may reduce the requirement for recurrent to the system.

6.3.7   In Situ Foam Fractionation

The removal of PFASs from groundwater has been suggested using an in situ downhole FF device (OPEC, 2017). Compressed air is used by the downhole system to concentrate PFASs into recoverable foam that is created within the well. Though carried out in situ, the idea is comparable to ozofractionation. Until now, this technique has not taken into account the possibility of obtaining a residue on Ignition (ROI) for air injection or foam concentrate collection outside of the well annulus. As a result, one possible worry is that this method will only remove PFASs from groundwater inside the well, leaving PFASs in groundwater outside the well annulus unaffected. There is a chance that PFAS-improved FF may form at the potentiometric surface of the groundwater if air were to leave the well and disperse radially since there is no way to collect FF outside of the well annulus. This FF loaded with PFAS may eventually re-influence groundwater. There are restrictions on potential advances that combine downhole FF with in-well groundwater circulation since the viability of in-well recirculation wells has been questioned. An analysis of groundwater in-well circulation technologies revealed that their limited use in groundwater remediation is due to short-circuiting issues and lower than anticipated return on investment (Allmon et al., 1999).

6.3.8   Reverse Osmosis and Nanofiltration

Regardless of chain length, RO and NF have effectively proven to be incredibly effective for the removal, treatment, remediation, or management of PFASs (Dickenson & Higgins, 2016) and are anticipated to be successful in eliminating a variety of PFA precursors. They are usually used in conjunction with extensive drinking water procedures, although they are costly. To avoid fouling or degradation of the RO/NF membrane in their applications to groundwater, the suspended particles and water geochemistry need to be evaluated, analyzed, and controlled. Because these methods do not remove PFASs, this strategy also creates a small volume, high CCN-rejected waste that has to be treated or disposed of.

6.4 DESTRUCTIVE TECHNIQUES

6.4.1 CHEMICAL OXIDATION

It has been widely demonstrated that chemical oxidation techniques are efficient in adapting PFA precursors to PFA compounds (Bruton & Sedlak, 2017); PFA carboxylates may be broken down by some oxidants, whereas PFA sulfonates provide serious obstacles to oxidative assault (Vecitis et al., 2009).

As reported by Schröder & Meesters (2005), innovative oxidation procedures that have an advanced oxidation potential of more than some physicochemical and biological reactions were unable to degrade PFOS at a CCN of 20.00 mg/L over the course of 7,200 seconds employing laboratory-scale experiments with O_3, O_3 hydrogen peroxide (H_2O_2), O_3/UV, and Fenton's reagent. Their finding was confirmed by another study using the same reagents to look at the PFA breakdown at the µg/L level (Qiu et al., 2006). Also, Hori et al. (2007) examined the photochemical breakdown of an FT unsaturated carboxylic acid in room temperature water, driven by persulfate.

With the use of a xenon mercury lamp to activate the persulfate, PFOA was completely degraded and very slightly formed short-chain PFA carboxylates. 40 micromolar radiation at a wavelength of 2.20×10^2 to 4.60×10^2 nanometres (nm), four hours of radiation, and 12.00 g/L persulfate were the circumstances that were observed (Hori et al., 2005). The FT molecule vanished entirely in five minutes, and after 180 minutes, 95.50% of F– and 104% of CO_2 were recovered (Hori et al., 2007). Additional research also showed that PFOA was efficiently destroyed by heat-activated persulfate (Hori et al., 2008). Microwave-induced persulfate has been used to characterize the degradation of PFOA. Under alkaline circumstances, it was shown that a lower pH led to a quicker breakdown and that a higher pH almost eliminated destruction (Lee et al., 2009).

According to research by Kerfoot (2013), O_3 in the form of nanobubbles can eliminate PFOS and PFOA. The huge surface area of the minute bubbles that formed, however, suggests that PFASs could possibly have been solution foam-fractionated-based because PFASs dispense to the gas–liquid interface and then moved to the surface of the liquid as foam in a procedure akin to ozo-fractionation. O_3-sparging PFASs to the water table and smear region may have caused the repartitioning of PFASs, which might have negatively impacted the usage of persulfate and O_3 documented for effective in situ treatment of PFASs (Eberle et al., 2017).

It has been reported that 89% of PFOA was destroyed in 150 minutes by utilizing H_2O_2 via Fenton's reagent with a molar H_2O_2 and 0.5 millimolar (mM) iron (III) (Mitchell et al., 2014). However, for these studies, there was no provision for a fluoride mass balance. PFASs may be removed from the aqueous phase via a sorptive process under low pH circumstances when a mineral precipitate, like iron, forms or if iron at the nanoscale is supplied. Subsequent experiments with H_2O_2 at pH 12.8 resulted in a 68% breakdown of PFOA in 7.20×10^3 seconds. According to Mitchell et al. (2014), near stoichiometric fluoride ion CCNs were found in these further experiments, indicating that PFOA had mineralized. As reported by Ahmad (2012), when H_2O_2 was catalyzed by iron (III), the rapid breakdown of PFOA was likewise observed, however, there was no provision for fluoride stoichiometry.

In summary, a number of publications describe the oxidative process based on radicals leading to the degradation of PFA carboxylates and FT compounds. Additionally, some of these reports demonstrate the equivalent stoichiometric production of fluoride. In contrast to PFA carboxylates, PFA sulfonates seem to be much more resistant to chemical oxidation. This might be the result of mechanistic reasons because, unlike with carboxylates, where carbon dioxide can form, there is hardly a stable and advantageous leaving group when the oxidative assault occurs on the sulfonate FG. The production of appreciable amounts of more transportable short-chain PFASs is one of the issues with chemical oxidation for in situ applications. Furthermore, taking into account reagent dispersion or subsurface heterogeneities, laboratory-based experiments are unable to duplicate reaction kinetics. Also, as it is evident when oxidizing NAPL, the presence of cationic and zwitterionic precursors in source region soils could possibly result in PFA rebound impacts.

6.4.2 Chemical Reduction

Owing to the extremely electronegative of the atoms of fluorine encircling the carbon skeleton to shield it from oxidative assault in the presence of hydroxyl radicals (•OH), common oxidative mechanisms may be unable to break down PFA compound. But since the fluorine atoms are so electronegative, they could be more vulnerable to reductive assault (Song et al., 2013). It has been shown that PFA compound can be effectively degraded by a number of advanced reductive processes (ARP), including UV-irradiated sulphite, -iodide, and -dithionite (Xuchun Li et al., 2014; Qu et al., 2010). To start the DFN process, the solvated electrons target the C–F bonds in the α-position rather than the carbon–carbon bonds (Song et al., 2013; Qu et al., 2010). Oxygen and anions, like nitrate, will be difficult to remove via ex situ water treatment, which will use up the reductants and drastically lower effectiveness. Strong and non-selective reductants, solvated electrons are usually produced by UV-irradiating reductants. Nevertheless, dissolved oxygen and nitrate readily scavenge them.

Also, there can be some worries about how treating water with sulphites or iodides would affect the ecosystem. While laboratory studies have proved variable efficiencies elimination for PFOA and PFOS using different ARP techniques, their practical application for in situ PFA remediation at the field scale is challenging. Further research is still required to determine the effectiveness of ARP in controlled settings and the role that solvated electron-based reduction plays in the mineralization of PFA.

6.4.3 Electrochemical Oxidation/Electrochemical Anodic Oxidation

One of the foremost evolving technologies is PFCA oxidation via electrochemical means (Zhuo et al., 2011). Direct transfer of electrons on the anode surface causes degradation, which makes it suitable for small-volume, high-CCN waste streams. But because short-chain PFASs exhibit lower efficacy and potential, concerns about corrosion of the electrode as well as the production of byproducts need to be taken into account (Merino et al., 2016). When treating or managing PFAS-polluted wastewater combined with co-pollutants, organics, chloride, or other hazardous compounds, noxious byproducts such as adsorbable organic halides, bromate, chlorine gas, fluoride, hydrogen, and perchlorate may be produced (Trautmann et al., 2015). There are two main ways that EO procedures are known to destroy contaminants: (1) Indirect EO: in this process, chemical oxidation occurs after powerful oxidants are generated on the anode. The $\bullet$OH produced by the anode is typically the result of this kind of indirect electro-oxidation; (2) Direct EO, in which the anode itself undergoes electro-oxidation directly via the production of physically adsorbed "active oxygen" (adsorbed $\bullet$OH). Numerous pollutants in wastewater can be efficiently mineralized by these techniques. According to Urtiaga et al. (2014), because membrane, AST, and/or ion-exchange procedures create waste that has to be managed, electro-oxidation provides an effective and sustainable alternative to these treatment and remediation technologies. Additional benefits of electrochemical deterioration are its automation simplicity, robustness, and adaptability (Anglada et al., 2009).

Gomez-Ruiz et al. (2017) conducted research on the electrochemical management, remediation, and treatment of eight PFASs at ecologically appropriate CCNs in industrial wastewater treatment plant effluent. The total PFAS CCN in the WWTP effluent was 1,652 mg/L. The main contributors to this content, which also included a sizeable quantity of short-chain PFCAs, were 6:2 FT sulfonamide alkylbetaine and 6:2 FT sulfonate, accounting for 92% of the total PFAS content. Ninety-nine point seventy per cent of detectable PFASs were removed when using a boron-doped diamond (BDD) anode. However, there was no evidence of the treatment and purification of PFA sulfonates, like PFOS and 40 mM (4 g/L) perchlorate formation was noted.

When using BDD for deterioration and mineralization, anodic material is essential. When it comes to leaching, tin- and lead-based electrodes have received less attention and have raised concerns. Aside from the problems with imperfections and pinholes in their composition, BDD is reliable and effective. To minimize the amount of defects, the diamond-film layer must be some-microns thick due to grain sizes ranging from 0.5 to 10 microns (Urtiaga et al., 2014). An improved alternative to conventional BDD is provided by ultrananocrystalline conductive BDD, which features a thin-film covering and nanoscale dimension.

The utilization of electrochemical remediation, management, and treatment for the breakdown of PFOA, PFOS, and other PFA compounds in AFFF-based groundwater taken from a former firefighter training location and PFA-spiked synthetic groundwater was evaluated in laboratory studies (Schaefer et al., 2015). With DFN verified for both PFOA and PFOS and 58% and 98% recovery as fluoride, respectively, (based upon the degraded mass of PFOA and PFOS), PFOA and PFOS breakdown were assessed utilizing a commercially made Ti/RuO2 anode. Also, treatment, remediation, and management of additional PFA compounds found in the groundwater were noted; shorter-chain PFA compounds were often more resistant to treatment.

It has been shown that both PFOS and PFOA may be efficiently degraded by using an electrode made of titanium suboxide (Ti_4O_7) (Huang, 2017). It has been claimed that PFOS degrades continuously and quickly on the Ti_4O_7 electrode, mineralizing to CO_2 and F^- while producing very little in the way of intermediate ONF chemicals. Due to the difficulties EO has previously encountered with PFA sulfonates, the annihilation of both PFOS and PFOA by EO using this Ti4O7 electrode sounded extremely promising. Since EO was previously used to treat industrial wastewater and produced 4.00 g/L of perchlorate, it seems that further study is needed before using this electrode to treat other PFAS waste streams in a commercial environment; as a result, byproduct generation needs to be evaluated (Gomez-Ruiz et al., 2017). For commercial use, electrochemical oxidation often has certain restrictions. If EO is appropriate for PFAS cleanup, more studies with environmental matrices are required to make that determination.

6.4.4 SONOLYSIS

To promote cavitation in water, sonolysis employs sound waves with frequencies typically ranging between 20 kilohertz (kHz) to 1.10×10^3 kHz. Cavitation is the simple process of a fluid forming microbubbles as a result of negative pressures. Cavitation is caused by cyclic cycles of rarefaction and compression in water caused by sound waves. Significant heat is generated as the microbubbles collapse during compression cycles, and findings from the literature indicate that temperatures of up to 5.00×10^3 ^{0}K may be reached inside the bubbles (Campbell et al., 2009). Numerous optimization considerations, including sound field circulation, bubbling gas in-line, pH adjustments, altering the outside temperature and pressure, and •OH production, can be investigated for scale-up. The effectiveness of these aspects may vary depending on the specific PFAS (Fernandez et al., 2016; Hao et al., 2014; Rodriguez-Freire et al., 2015).

Greater frequency sonolysis produces smaller bubbles (more surface area) with lower energy production, whereas lesser frequency sonolysis produces larger bubbles with better energy output (Drees, 2005). For input energy reasons, the majority of non-PFA compound sonolysis applications range between 20 kHz to 40 kHz. Based on findings in the literature, a frequency ranging from 5.00×10^2 kHz to 1.10×10^3 kHz is suitable to remediate a wide spectrum of organic pollutants (including PFA compound) (Rodriguez-Freire et al., 2015, 2016, 2020; Fernandez et al., 2016), however, it is advised to specify a design frequency using site-specific, real-time data observations. Some studies have demonstrated the use of sonolysis for the treatment, remediation, and management of PFAS at frequencies greater than 2.00×10^2 kHz in order to optimize the surface area as well as the resulting interaction between the molecules of PFASs and microbubbles (Hao et al., 2014; Rodriguez-Freire et al., 2015). Since heat breakdown mediated at or inside the bubble surface is the predominant treatment, remediation and management mechanism associated with sonolysis for PFAS, AST onto or interaction with the surface of these microbubbles is crucial to the sonolytic treatment, remediation, and management of PFAS. A specific PFAS's hydrophilic FG, such as a carboxylate or sulfonate group, is still preferentially soluble in the liquid phase, while the fluorine-saturated carbon chains within it are preferentially drawn to the gas phase. Thus, the gas–liquid interface of a submerged bubble is perfect for PFAS accumulation, and it makes sense to use higher frequency (HF) sonolysis to maximize the microbubbles' accessible surface area.

With steady observations of the rate of the pseudo-first order kinetics as well as the quicker kinetics for larger PFASs with more fluorination (perfluorinated > polyfluorinated), sonolysis seems to extinguish several varieties of PFAS composites (both the long chain as well as the short chain) (Fernandez et al., 2016; Rodriguez-Freire et al., 2016). One of the weaknesses of the sonolysis data available for PFASs is that large doses of PFAS (> 10,000 ng/L) are employed since viability is the main concern. Evidence exists to support the appropriate management of common competing pollutants and groundwater geochemistry and could possibly provide a few minor advantages (Rodriguez-Freire et al., 2015). Finally, PFAS content in water might be used as a pretreatment technique to maximize sonolysis. Because sonolysis is limited to the liquid stage, PFASs adsorbed to solids would need to be leached using extractants in order to facilitate sonochemical destruction in the liquid phase.

In the chemical industry, sonolysis is widely used for material processing and solution mixing, usually on a small scale with tiny quantities (unlike vast volumes of water in public systems). By fracturing the sludge composites and boosting anaerobic digestion, also, it has also been used in processing biological sludge for the production of more biogas. The technique provides modular response unit architecture for upscaling. A transducer, which also focuses the energy, produces sound and transmits it to water. Commercial transducers come in a variety of forms, and choosing one relies on a number of variables, such as the needed energy intensity, frequency, and reactor dimension (size and shape). In order to achieve uniform cavitation, the quantity of modular units or transducers required in a big tank will rely not only on the flow rate and reaction kinetics to be purified or treated but as well as on the attainable sound field.

There are several scholarly publications available about the sonolysis-induced breakdown of a broad range of chemical substances. Hydrogen fluoride (HF) ultrasound has been found to be beneficial for the degradation of PFASs. Although the energy expenses for operation are reasonable (0.1 to 0.3 kWh per litre of treated or purified water), the units' capital cost usually restricts the range of applications.

At the moment, sonolysis is yet to be scaled up for the commercial application, but it has been shown at the laboratory scale for PFASs. Given the restricted zones of efficacy for cavitation bubble propagation from transducers, scaling up is probably going to provide considerable design issues. A structure that encourages a homogeneous distribution of cavitational activity is crucial to the design of a sonochemical reactor. The quantity and placement of transducers, frequency, reactor dimension, as well as dissipation of power, all affect this kind of activity. Larger sonolytic reactors may also have dead zones (Gole et al., 2018).

6.5 SOIL AND SEDIMENT REMEDIATION, TREATMENT, OR MANAGEMENT OF PFAS

The many remediation, treatment, or management technologies that are important to per- and multifunctional agricultural sprays (PFASs) have been grouped on a scatter plot based on their practicality and state of development (Figure 6.3). When it comes

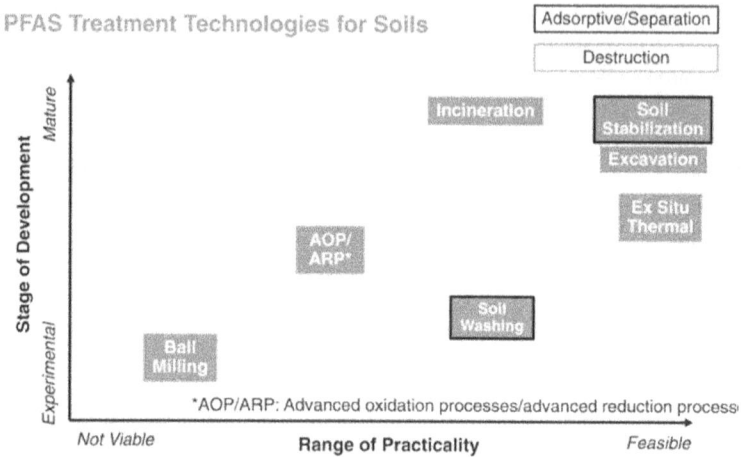

FIGURE 6.3 A scatter plot of PFAS the treatment, remediation, or management technologies for soil and sediment.

to PFAS soil remediation, the following standard techniques are applicable: excavation combined with off-site disposal in a landfill or incineration, covering or capping while keeping an eye on precipitation infiltration, and soil washing.

For PFAS-impacted source zones, excavation with offsite removal in a landfill is important, however, given PFAS tenacity and the inadequate PFAS remediation, treatment, management, or monitoring in several landfill leachates, this option should be carefully assessed for potential long-term responsibility in addition to expense. Landfill operators are being more stringent when it comes to trash touched by PFAS in a number of nations, most notably Sweden and Australia. To eliminate PFASs, excavated soils can be burned at temperatures above 1,100 °C, although this may be too costly for many locations. Wastes that include more than 50 milligrams of per kilogram of PFOS (permanent organic pollutant) in the UK may need to be destroyed, even if they are categorized as non-hazardous. Long-term management is needed for both the current practices of capping soil effects left in place and confining excavated dirt inside designed stockpiles to stop infiltration and leaching to groundwater. Redevelopment limitations and ongoing obligations are important factors for this management strategy. To reduce the amount of PFAS-impacted soil, soil washing or actively leaching PFAS from soil composites ex situ for the collection of the PFAS-rich leachate, might be an appropriate strategy. Due to the potential complexity and expense of leachate treatment, remediation, or management and finest treatment, remediation, or management/disposal, appropriate PFAS-based soils usually have comparatively little fine contents.

Some of the remaining PFAS-relevant soil treatment, remediation, or management technologies are mostly destructive (ARP or thermal treatment, remediation, or management) or stabilizing (fixation) of the soil, and they are covered in the subsections that follow.

6.5.1 STABILIZATION

Groundwater may be impacted by PFAS for decades due to shallow soil and aquifers under FTAs. At present, the state of art for managing PFAS source zones involves physical confiscation by excavation or extraction combined with above-ground treatment, remediation, or management systems. An alternative to ex situ treatment, remediation, or management of PFAS wastes is in situ stabilization (ISS) of PFAS utilizing in situ soil mixing (ISM) with adsorbents, which is meant to shield groundwater from PFAS leakage in the future. Due to its ability to homogenize geological anisotropy and provide instant access for soluble PFAS contained in low-porous layers, ISM is a useful application approach for reaching source material. The ability to treat porewater and soil in both the vadose and saturated zones is another benefit of ISM. Reducing the source zones' long-term leaching potential is one way that using ISM in conjunction with adsorbents might reduce environmental risk.

Although the technique applies to FTA source regions, it is crucial to comprehend the stable durability because PFASs are not eliminated. Considerable work has been done in the last few years to discover efficient adsorbents and comprehend the AST processes of different PFASs. AC, organo-modified clays, and unique mixes of AC, clay, and aluminium hydroxides are examples of commercially available goods that have been suggested and tested for the International Space Station (Du et al., 2014). Iron oxides, layered double hydroxides, and graphene derivatives are among the materials being developed for PFAS stabilization (Hu et al., 2017; Lath et al., 2017; Pennell et al., 2017). But in order to confirm that ISS is a suitable PFAS cleanup technique, long-term leachability testing conducted under "worst case" environmental scenarios is necessary. Although there are efforts in 2018 aimed at doing this, the long-term persistence of fixation has not yet been shown in practice. For instance, the authors of this paper and the US Air Force will work together to develop an ISM program at the field level in an FTA source zone located in the southwest region of the country. Using three commercially available adsorbents, this initiative will stabilize PFASs in an FTA source region and track leachability over a three- to four-year period.

It has been reported that commercial items perform well in leaching testing that utilizes derivative tests or the acidic toxicity characteristic leaching process (Ziltek, 2017). These assays, however, were intended to evaluate the leaching of cationic metals in circumstances of small pH value that assist in the promotion of more destructive leaching. Because the small pH causes the dominating charges on sorbent composites to convert to positive charges, which facilitates the attachment of the anionic PFA compound to the sorbent, the low pH offers considerably fewer hostile conditions for the anionic PFA compound. Since the pH of most sites (apart from landfills) is expected to be high or circumneutral, it seems unsuitable to apply the acidic leaching experiments. The tests were created for acidic landfills, and if sorbents are employed for ISM at locations where acidic leach tests are offered to prove performance, there are worries that they would not receive realistic test results. A 24-hour shake extraction at a pH value of 7 (application rates ranging from 5% to 30%) is one example of a leach test at neutral pH that has been published. However, the only data demonstrating reduced leaching came from a 25% application rate, which is far too high to be practical and economical for commercial purposes (Braunig et al., 2017). It appears that more research is needed to prove the efficacy of novel sorbents and commercial items using suitable leach tests.

Further research will examine the influence of pozzolanic chemistry on the efficient stabilization of PFAS. Portland cement chemistry, such as its alkaline pH, has to be geochemically evaluated for its impact on PFAS AST since Portland cement is frequently used in conjunction with ISM to increase the treated soil's unconfined compressive strength. A sustainable method that avoids producing concentrated waste that needs to be managed or destroyed elsewhere is managing PFAS source zone soils in situ.

6.5.2   HIGH ENERGY ELECTRON BEAM

High energy eBeams, or EBs, are produced from electricity using electron accelerators and are high-efficiency, non-thermal, flow-through, chemical-free techniques (Cleland, 2012; Pillai & Shayanfar, 2017). According to Zembouai et al. (2016), Pillai (2016), and Pillai & Shayanfar (2017), global commercialization of the technique has occurred for pasteurizing feeds, sterilization of medical equipment, cross-linking polymers, as well as for getting rid of bugs and vermin from fresh yields. It offers an alternative to using radioactive isotopes for ionizing irradiation. Dose is the quantity of eBeam energy absorbed per unit mass by an irradiated substance. The kind and thinness of the composites, the strength of the beam, as well as the duration of the material's exposure to the EB all affect the absorbed dosage during eBeam therapy (Waite et al., 1998).

eBeam may be used on both liquid and soil matrices for a variety of reasons. Cleaning up sewage sludge (Praveen et al., 2013); remediation or treatment of soils polluted with heavy hydrocarbons (Briggs & Staack, 2015); cleanup of semi-volatile and volatile organic substances found in liquid wastes including landfill leachate, wastewater, and groundwater (US.EPA, 1997, 2024a, b). Three main reactive species are created when water is exposed to radiation: strong oxidizing and reducing hydrogen radicals and solvated electrons; also, strong reducing •OH is produced. In doing so, sophisticated reduction and oxidation procedures are produced without the need for additional substances. In potable, raw, and secondary wastewater effluent, the absolute CCN of radicals generated during irradiation has been observed at more than mM levels, however, this CCN is dose- and water-quality dependent (Waite et al., 1998).

Researchers from Texas A&M University have used eBeam technology to show the DFN of PFOA in aqueous samples (Wang et al., 2016). The molar CCN of free fluoride ions as well as the starting molar CCN of PFOA to be treated or purified were used to calculate the DFN efficiency in the study. Final DFN efficiency varied with increasing nitrate, alkalinity, and fluvic acid CCNs, ranging from 34.60% to 95%. The production of secondary radicals and aqueous electrons may be the cause of the DFN (Wang et al., 2016). In an anoxic alkaline solution (pH = 13), another research further confirmed EB-mediated DFN of PFOS and PFOA with disintegration efficiencies of 95.70% for PFOA and 85.90% for PFOS. The aqueous electron and hydrogen radical were shown to be significant in the EB dilapidation of PFOA and PFOS, according to investigations using radical scavenging (Ma et al., 2017). In order to fully understand treatment, remediation or management performance potential and detect any harmful byproducts, additional testing over a range of CCNs and assessment of this technology for treating and managing other PFASs (polyfluorinated precursors as well as other long- and short-chain PFA compound) in terrestrial and aquatic environment will be required.

6.5.3 Low/High-Temperature Thermal Desorption

Widespread use of this method has been made to remediate pesticides in soils that share PFA-like physicochemical characteristics. PFASs may volatilize in the oven between 400 and 500 degrees Celsius, maybe with an off-gas treatment between 900 and 1,000 degrees Celsius. The primary issue with thermal treatment is how well it removes PFASs from soils. According to preliminary results, PFOS remediation, treatment, or management in more severely affected soils is unproductive, even at 6.00×10^2 °C for 60 mins of residence time (Fisher 2017). This is one of the main concerns with thermal treatment. Concerning this method is also its capacity to volatilize the precursors with greater molecular weights, which are perhaps more volatile than the PFA compound. Emissions of hydrofluoric acid and other fluoro-organics from the treatment or remediation system are additional factors to take into account.

In large ex situ treatment plants, thermal desorption (TD) for PFASs entails heating dug soil to a temperature of about 500 to 600 °C employing a rotating kiln equipped with thermal screws or gas burners to desorb PFASs into the gas stream. At a temperature greater than 1,000 °C, the afterburner catalyzes the oxidation of PFASs. Although TD appears to be a potentially workable strategy for soils affected by PFAS, no full-scale application has been made that expressly targets PFASs, and there is presently no performance data on the effectiveness of polyfluorinated precursor remediation or treatment available. Study and advancement for TD are currently focused on improving performance via temperature tuning, efficient vapour management or treatment (e.g., using air stream catalytic oxidation, treating condensate water), and problems with vapour-scrubbing to eliminate hydrofluoric acid and other produced byproducts.

When assessing TD, it is important to take into account the mobilization cost and related output rate because rotary kilns for TD can be rather expensive. For lower treatment volumes, on-site TD might not be financially feasible. Additionally, pretreatment and/or longer treatment durations may be necessary for less cohesive soils, which might affect the technology's reasonable price point. Although soil may be moved to a permanent TD plant, high-temperature incineration would seem to be preferred due to comparable transportation and disposal costs as well as less confidence around TD.

TD can also be accomplished by employing thermopiles. Excavated soil is placed into enclosed piles using thermopiles that are heated using heater rods or gas/diesel burners to desorb PFASs into the vapour stream (5.00×10^2 °C to 6.00×10^2

°C necessary). In order to recover vapours that are then treated with thermal oxi-dizers or condensers to eliminate PFASs, thermopiles are enclosed and held under vacuum. Depending on the kind of soil, moisture content, and degree and kind of PFAS adulteration, raised soil temperatures must be preserved for some weeks in order to accomplish effective management or treatment. However, thermopile-based TD has not yet been reportedly used in soil affected by PFAS. It is questionable if the 5.00×10^2 °C to 6.00×10^2 °C required for PFAS desorption can be reached and maintained because these temperatures are significantly higher than what is usually attained in thermopiles. The operation of a thermal waste recovery capability as well as the accompanying off-gasses will probably need a high level of licensing and stakeholder participation.

6.5.4 VAPOUR ENERGY GENERATOR PROCESS (VEG)

In an ex situ treatment chamber, the VEG procedure employs steam heated to 1.10×10^3 °C for the elimination of PFASs from damaged soils. By burning the synthesis gas (syngas) produced by splitting water and utilizing the carbon monoxide produced by heating the soil's organic portion, it produces extra heat. Compared to large-scale TD systems, this technology is thought to have several advantages, including as reduced energy costs, a smaller operational footprint, and a cheaper mobilization cost. All PFASs should be eliminated by using steam at 1.10×10^3 °C. A full-scale VEG deployment for the cleanup of PFASs in soil is also suggested, along with a thorough small-scale experiment scheduled for the first and second quarters of 2018 in California for a site in Europe.

Endpoint Consulting Inc. (San Francisco, CA) has patented a small, very efficient steam generator that is used in the VEG technique, an ex situ TD and destruction method. It is possible to reach soil temperatures of up to 9.50×10^2 °C during VEG treatment, at which point PFASs are destroyed and desorbed into the vapour phase. About 45 full-scale projects have been accomplished in the US using the VEG tech-nology for improved oil recovery (both in situ and ex situ remediation) for a variety of recalcitrant pollutants, including pesticides, heavy-end oils, hydrocarbons from petroleum, polychlorinated biphenyls, polycyclic aromatic hydrocarbons, as well as some selected metals (As, Zn, and Hg). The authors are aware that research efforts aimed at treating PFAS-containing wastes and wasted adsorbents are also getting underway.

VEG uses a multistep vapour treatment method that includes steam, caustic, lime, and zero-valent iron. The VEG unit cycles back any desorbed PFASs so they can thermally deteriorate at temperatures higher than 1,100 °C. Scrubbers made of car-bon dioxide are used to cut greenhouse gas emissions. Caustic is also used to cleanse acid gases (hydrofluoric acid, hydrochloric acid, etc.). The water treatment procedure could possibly produce syngas that lowers the need for propane in the steam genera-tor, which normally runs on air, recycled water, and propane. Crucially, in a closed loop system, treated or purified vapours, condensed water, as well as syngas are all returned to the VEG unit with negligible or no vapour emissions.

When it comes to execution, the VEG technology may be used with a mobile procedure that can move locations more quickly than TD units and even smaller batch procedures are obtainable if needed. Although the residence period within the VEG has a significant impact on the normal throughput rate, it usually achieves 200 cubic meters per day. Although at a slow rate, the procedure produces concentrated salt liquid waste (such as bisulfates, chloride, fluorides, and sodium nitrates).

In 1997, Endpoint Consulting and the Colorado School of Mines collaborated to conduct a bench-scale investigation (Endpoint, 1997) to evaluate the remediation of AFFF-contaminated soils. With a 30-min residence period at 9.50×10^2 °C, the tests demonstrated > 99% removal of PFA compound, together with PFOA, PFOS, as well as PFHxS. The study also showed that acid gases might be effectively removed. It did not, however, evaluate polyfluorinated precursors; further investigation into the fate of PFASs within the system is necessary.

Although there has not been a reported full-scale implementation of VEG expressly to treat PFASs, auspicious small-scale investigations have been carried out, and the technology has been used at many full-scale developments for pollutants other than PFASs.

6.5.5 Ball Milling (BM)

Using stainless steel balls with sizes ranging between 5 to 10 mm, BM is a destruction-based technology that is used in traditional planetary ball mills. Throughout the BM process, several collisions between the deformable solid phase waste being processed and the non-deformable steel balls occur as a result of the balls and the solid phase dirt being spun at several hundreds of revolutions per minute with directional variations. The temperature of the solid phase waste surface rises as a result of the deformation at its surface. BM is a type of mechano–chemical destruction, which describes reactions that take place at chemical surfaces when a mechanical force is applied as a result of a brief rise in temperature or the production of triboplasmas (i.e., highly ionized neutral gas; Heinicke & Hennig, 1984). To take advantage of the possibility of producing the •OH and enable simultaneous chemical destruction, co-milling agents like potassium hydroxide (KOH), lime (CaO), silicon dioxide (SiO_2), and sodium hydroxide (NaOH) may be included in various BM procedures. As evidenced by X-ray diffraction, a BM example for PFOA and PFOS in the literature used the accumulation of KOH to show how the PFA compound was destroyed by the BM process, producing potassium sulfate (K_2SO_4) and fluoride (KF) (Zhang et al., 2013).

More than 90% of PFOA and PFOS were shown to decrease in the research, while more than 95% of fluoride and sulfate were recovered (Zhang et al., 2013). There are several reasons to think that thermal destruction, not chemical destruction, was the real dominating destruction mechanism for PFOA and PFOS, even though Zhang et al. (2013) ascribed part of the destruction to the presence of the co-milling agent and some type of chemical treatment. The feasibility of BM in relation to the issues caused by PFASs has not been thoroughly investigated, However, if thermal destruction is the primary mechanism, then, providing it can reach the temperature required for PFAS-specific thermal destruction, it should apply to a broad spectrum of PFASs. The feasibility of deploying planetary ball mills to PFAS-impacted sites varies based on the quantity of soil to be processed and an evaluation of the cost-benefit ratio in relation to off-site incineration. To enable a suitable soil throughout and related output rate, the planetary BM would require to be sized properly. Bench-scale testing could establish the final time needed to accomplish total destruction using BM for PFOA and PFOS, this time has been found to be higher than 90% after 66 mins (Zhang et al., 2013).

6.6 CONCLUSION AND RECOMMENDATIONS FOR THE REMEDIATION, TREATMENT, OR MANAGEMENT OF PFAS

Remediation, treatment, and management of PFAS employing specific passive adsorptive/filtering or active degradative/destructive procedures may reach a promising remediation, treatment, and management efficiency, according to a comprehensive scientific review. However, according to Lu et al. (2020), those published single-method studies have significant shortcomings, such as:

- In order to treat the wasted adsorbent and concentrated retentate, ex situ PFAS removal, remediation, treatment, and management technologies like GAC/IXR AST and NF must use an extra step (like burning).
- High acidity or alkalinity, high processing temperatures, and extensive usage of H_2O_2 and/or zero-valent iron are typical harsh operating conditions needed for in situ degradation of PFAS employing a single advanced oxidation method.
- During inadequate mineralization, shorter-chained intermediates that are produced from the breakdown of the original PFAS may release or volatile escape, potentially leading to significant secondary contamination.

The studies that were selected utilizing treatment, remediation, treatment, and management train processes in this chapter are divided into two primary approaches: (1) combining an adsorptive/filtering method with a degradative/destructive technology in tandem and (2) combining several degradative/destructive technologies in parallel. It is believed that the parallel combination of electro-Fenton (EF) and EO and the tandem connection of NF with EO (NF-EO) are the most remarkable systems in the various groups among the findings from these studies. The primary reasons for this are:

- Industrial wastewater with a comparatively high CCN of PFAS (64 and 204 mg/L) could be treated directly by the NF-EO system.
- The degradation kinetic rate of the NF-EO system was 2.25 h^{-1}.
- The degradation of PFHxA ($C_5F_{11}COOH$), which requires more energy to penetrate the CeC bonds because of the stability from a shorter carbon chain, was evaluated using the NF-EO system.

- The NF-EO system is a workable strategy that is simple to apply at an already-existing treatment, remediation, treatment, or management centre.
- The EF/EO system successfully integrates two related advanced oxidation processes, which employ the same input energy (electric).
- To maximize the conversion of electric energy to chemical energy, the EF/EO system can use the anode and cathode to concurrently form reactive radicals.
- To lower the main chemical input, the EF/EO system may naturally synthesize H_2O_2, which is an essential chemical reagent for producing reactive radicals. With just slight adjustments to the cathode's composition, the EF/EO system can be set up using the standard EO system.

It is crucial to continue developing treatment, remediation, treatment, and management technologies and lowering the overall cost of chemicals and energy consumption in order to support the treatment, remediation, treatment, and management training strategy for future real-world applications in PFAS cleanup. The best technical approach might be to further integrate the NF-EO (tandem treatment train) and EF/EO (parallel treatment train) systems into a new system that combines EO, EF degradation, and (NF-EO/EF), because it might greatly reduce the need for H_2O_2 and treat PFAS in a wide range of CCNs. Furthermore, portable solar panels as promised by Lu et al. (2020), as illustrated in Figure 6.4, as adopted from Lu et al.

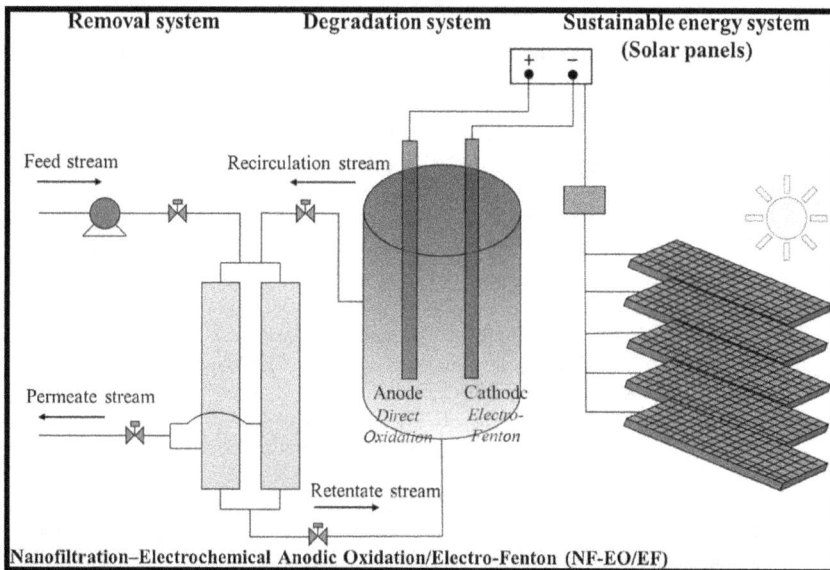

FIGURE 6.4 Diagrammatic representation of a treatment, remediation, and management train system (NF-EO/EF) linked to a solar power system for sustainable in situ treatment, remediation, and management of PFAS. Adapted and reproduced from Lu et al. (2020) with permission from Elsevier B.V.

(2020), could generate the electricity needed for the circulation system and electro-chemical procedures, potentially ensuring that the entire system ultimately becomes self-sufficient in terms of both chemical and energy usage.

In conclusion, a combination of several removal, remediation, treatment, management and/or degrading techniques can limit the creation of hazardous byproducts, lower overall costs, and increase removal, remediation, treatment, and management efficiency. The most practical PFAS remediation, treatment, and management of technology must ultimately be chosen after a thorough field study of the local hydro-geology and climate, the extent of pollution, co-contaminants, and neighbouring communities. Considering all of these factors, it is anticipated that PFAS removal, remediation, treatment, and management will be more difficult than traditional pump-and-treat or soil-vapour-extraction methods. It is believed that in situ treat-ment, remediation, and management of PFAS using treatment, management, and remediation train procedures will eventually emerge as a viable option for large-scale remediation, treatment, and management projects as cost-effectiveness and treatment, remediation, and management of technologies continue to advance.

REFERENCES

Ahmad, M. (2012). *Innovative oxidation pathways for the treatment of traditional and emerging contaminants.* Washington State University.

Ahrens, L., Xie, Z., & Ebinghaus, R. (2010). Distribution of perfluoroalkyl compounds in seawater from Northern Europe, Atlantic Ocean, and Southern Ocean. *Chemosphere, 78*(8), 1011–1101.

Allmon, W. E., Everett, L. G., Lightner, A. T., Alleman, B., Boyd, T. J., & Spargo, B. J. (1999). *Groundwater circulating well technology assessment.* Rep. No. NRL/PU/6115-99, 384. Naval Research Laboratory.

Anglada, Á., Urtiaga, A., Ortiz, I. (2009). Contributions of electrochemical oxidation to waste-water treatment: fundamentals and review of applications. Journal of Chemical Technology & Biotechnology, 84(12), 1747-1755

Appleman, T. D., Higgins, C. P., Quiñones, O., Vanderford, B. J., Kolstad, C., Zeigler-Holady, J. C., & Dickenson, E. R. V. (2014). Treatment of poly-and perfluoroalkyl substances in US full-scale water treatment systems. *Water Research, 51*, 246–255.

Backe, W. J., Day, T. C., & Field, J. A. (2013). Zwitterionic, cationic, and anionic fluori-nated chemicals in aqueous film forming foam formulations and groundwater from US military bases by nonaqueous large-volume injection HPLC-MS/MS. *Environmental Science & Technology, 47*(10), 5226–5234.

Benskin, J. P., Ikonomou, M. G., Gobas, F. A. P. C., Begley, T. H., Woudneh, M. B., & Cosgrove, J. R. (2013). Biodegradation of N-ethyl perfluorooctane sulfonamido etha-nol (EtFOSE) and EtFOSE-based phosphate diester (SAmPAP diester) in marine sedi-ments. *Environmental Science & Technology, 47*(3), 1381–1389.

Blaine, A. C., Rich, C. D., Hundal, L. S., Lau, C., Mills, M. A., Harris, K. M., & Higgins, C. P. (2013). Uptake of perfluoroalkyl acids into edible crops via land applied biosolids: Field and greenhouse studies. *Environmental Science & Technology, 47*(24), 14062–14069.

Blaine, A. C., Rich, C. D., Sedlacko, E. M., Hundal, L. S., Kumar, K., Lau, C., Mills, M. A., Harris, K. M., & Higgins, C. P. (2014a). Perfluoroalkyl acid distribution in various plant compartments of edible crops grown in biosolids-amended soils. *Environmental Science & Technology, 48*(14), 7858–7865.

Blaine, A. C., Rich, C. D., Sedlacko, E. M., Hyland, K. C., Stushnoff, C., Dickenson, E. R.
    V., & Higgins, C. P. (2014b). Perfluoroalkyl acid uptake in lettuce (Lactuca sativa) and
    strawberry (Fragaria ananassa) irrigated with reclaimed water. *Environmental Science
    & Technology*, *48*(24), 14361–14368.
Braunig, J., Baduel, C., & Muller, J. (2017). Influence of a commercial sorbent on the leach-
    ing behaviour and bioavailability of selected perfluoroalkyl acids (PFAAs) from soil
    impacted by AFFF. *Proceedings of Dioxin 2017 Conference*, Vancouver, Canada.
Bruton, T. A., & Sedlak, D. L. (2017). Treatment of aqueous film-forming foam by heat-
    activated persulfate under conditions representative of in situ chemical oxidation.
    Environmental Science & Technology, *51*(23), 13878–13885.
Campbell, T. Y., Vecitis, C. D., Mader, B. T., & Hoffmann, M. R. (2009). Perfluorinated
    surfactant chain-length effects on sonochemical kinetics. *The Journal of Physical
    Chemistry A*, *113*(36), 9834–9842.
Chen, X., Xia, X., Wang, X., Qiao, J., & Chen, H. (2011). A comparative study on sorption of
    perfluorooctane sulfonate (PFOS) by chars, ash and carbon nanotubes. *Chemosphere*,
    83(10), 1313–1319.
Choyke, B.-H. K. A. R. S. C., Riddell, S. O. K. M. A., & Higgins, N. M. R. F. P. L. (2017).
    CP Field JA Discovery of 40 classes of per-and polyfluoroalkyl substances in his-
    torical aqueous film-forming foams (AFFFs) and AFFF-impacted groundwater.
    Environmental Science & Technology, *51*, 2047–2057.
Cleland, M. R. (2012). Electron beam materials irradiators. In R. W. Hamm & M. E. Hamm
    (Eds.), *Industrial accelerators and their applications* (pp. 87–137). World Scientific.
Colosi, L. M., Pinto, R. A., Huang, Q., & Weber, W. J. J. (2009). Peroxidase-mediated deg-
    radation of perfluorooctanoic acid. *Environmental Toxicology and Chemistry: An
    International Journal*, *28*(2), 264–271.
Cornelsen. (2017). The PFC challenge. https://www.cornelsen.co.uk/perfluorad-pfc-treatment/
CSWRCB. (2019). *California state water resources control board-perfluorooctanoic acid
    (PFOA) and perfluorooctanesulfonic acid (PFOS)*. Division of Drinking Water. https://
    www.waterboards.ca.gov/drinking_water/certlic/drinkingwater/PFOA_PFOS.html
D'Agostino, L. A., & Mabury, S. A. (2017). Aerobic biodegradation of 2 fluorotelomer sul-
    fonamide–based aqueous film–forming foam components produces perfluoroalkyl car-
    boxylates. *Environmental Toxicology and Chemistry*, *36*(8), 2012–2021.
Dasu, K., & Lee, L. S. (2016). Aerobic biodegradation of toluene-2, 4-di (8: 2 fluorotelomer
    urethane) and hexamethylene-1, 6-di (8: 2 fluorotelomer urethane) monomers in soils.
    Chemosphere, *144*, 2482–2488.
Dasu, K., Lee, L. S., Turco, R. F., & Nies, L. F. (2013). Aerobic biodegradation of 8: 2 flu-
    orotelomer stearate monoester and 8: 2 fluorotelomer citrate triester in forest soil.
    Chemosphere, *91*(3), 399–405.
Dasu, K., Liu, J., & Lee, L. S. (2012). Aerobic soil biodegradation of 8: 2 fluorotelomer stea-
    rate monoester. *Environmental Science & Technology*, *46*(7), 3831–3836.
Dauchy, X., Boiteux, V., Bach, C., Colin, A., Hemard, J., Rosin, C., & Munoz, J.-F. (2017).
    Mass flows and fate of per-and polyfluoroalkyl substances (PFASs) in the wastewa-
    ter treatment plant of a fluorochemical manufacturing facility. *Science of the Total
    Environment*, *576*, 549–558.
Dawson, D. E., Lau, C., Pradeep, P., Sayre, R. R., Judson, R. S., Tornero-Velez, R., &
    Wambaugh, J. F. (2023). A machine learning model to estimate toxicokinetic half-lives
    of per-and polyfluoro-alkyl substances (PFAS) in multiple species. *Toxics*, *11*(2), 98.
    https://doi.org/10.3390/toxics11020098
Deng, S., Niu, L., Bei, Y., Wang, B., Huang, J., & Yu, G. (2013). Adsorption of perfluorinated
    compounds on aminated rice husk prepared by atom transfer radical polymerization.
    Chemosphere, *91*(2), 124–130.

Deng, S., Zhang, Q., Nie, Y., Wei, H., Wang, B., Huang, J., Yu, G., & Xing, B. (2012). Sorption mechanisms of perfluorinated compounds on carbon nanotubes. *Environmental Pollution*, *168*, 138–144.

Dickenson, E. R. V, & Higgins, C. (2016). Treatment mitigation strategies for poly- and perfluoroalkyl substances [Project# 4322]. Retrieved from Denver, Colorado. http://www.Waterrf.Org/PublicReportLibrary/4322.Pdf

Dickson, M. D. (2013). Method for treating industrial waste. https://pericles.ipaustralia.gov.au/ols/auspat/applicationDetails.do?applicationNo=2012289835

Dickson, M. D. (2014). United States US 2014O190896A1 Patent Application Publication. https://patentimages.storage.googleapis.com/a4/4f/ 92/6864036f22203d/US20140190896A1.pdf.

Dinglasan, M. J. A., Ye, Y., Edwards, E. A., & Mabury, S. A. (2004). Fluorotelomer alcohol biodegradation yields poly-and perfluorinated acids. *Environmental Science & Technology*, *38*(10), 2857–2864.

Dow. (2016). *Dowex Optipore V493 and L493 Resin*. https://www.lenntech.com/products/Dowex-Resins/Dowex-Optipore-L493-and-V493/Dowex-Optipore-L493-and-V493/index.html

Drees, C. W. (2005). *Sonochemical degradation of perfluorooctane sulfonate (PFOS)*. The Ohio State University.

Du, Z., Deng, S., Bei, Y., Huang, Q., Wang, B., Huang, J., & Yu, G. (2014). Adsorption behavior and mechanism of perfluorinated compounds on various adsorbents—A review. *Journal of Hazardous Materials*, *274*, 443–454.

Eberle, D., Ball, R., & Boving, T. B. (2017). Impact of ISCO treatment on PFAA co-contaminants at a former fire training area. *Environmental Science & Technology*, *51*(9), 5127–5136.

Endpoint. (1997). *Bench-scale VEG research & development study: Implementation memorandum for ex-situ thermal desorption of perfluoroalkyl compounds (PFCs) in soils*. Technical Note. Retrieved From.

EPA. (2016). *Fact Sheet PFOA & PFOS drinking water health advisories (EPA 800-F-16-003)*. Washington, DC, USA. https://www.epa.gov/ground-water-and-drinkingwater/supporting-documents-drinking-water-health-advisories-pfoa-and-pfos

Eschauzier, C., Beerendonk, E., Scholte-Veenendaal, P., & De Voogt, P. (2012). Impact of treatment processes on the removal of perfluoroalkyl acids from the drinking water production chain. *Environmental Science & Technology*, *46*(3), 1708–1715.

Evocra. (2016). OCRA process overview. https://evocra.com.au/about-us/our-story

Fernandez, N. A., Rodriguez-Freire, L., Keswani, M., & Sierra-Alvarez, R. (2016). Effect of chemical structure on the sonochemical degradation of perfluoroalkyl and polyfluoroalkyl substances (PFASs). *Environmental Science: Water Research & Technology*, *2*(6), 975–983.

Fisher, R. M., Le-Minh, N., Sivret, E. C., Alvarez-Gaitan, J. P., Moore, S. J., & Stuetz, R. M. (2017). Distribution and sensorial relevance of volatile organic compounds emitted throughout wastewater biosolids processing. *Science of The Total Environment*, 599–600, 663-670. https://doi.org/10.1016/j.scitotenv.2017.04.129

Gao, D., Du, L., Yang, J., Wu, W.-M., & Liang, H. (2010). A critical review of the application of white rot fungus to environmental pollution control. *Critical Reviews in Biotechnology*, *30*(1), 70–77.

Gao, K., Chen, Y., Xue, Q., Fu, J., Fu, K., Fu, J., Zhang, A., Cai, Z., & Jiang, G. (2020). Trends and perspectives in per-and polyfluorinated alkyl substances (PFASs) determination: Faster and broader. *TrAC Trends in Analytical Chemistry*, *133*, 116114.

Gao, K., Fu, J., Xue, Q., Li, Y., Liang, Y., Pan, Y., Zhang, A., & Jiang, G. (2018). An integrated method for simultaneously determining 10 classes of per-and polyfluoroalkyl substances in one drop of human serum. *Analytica Chimica Acta*, *999*, 76–86.

Garg, A., Shetti, N. P., Basu, S., Nadagouda, M. N., Tejraj M., & Aminabhavi, T. N. (2023). Treatment technologies for removal of per- and polyfluoroalkyl substances (PFAS) in biosolids. *Chemical Engineering Journal*, *453*, 139964.

Glüge, J., Scheringer, M., Cousins, I. T., DeWitt, J. C., Goldenman, G., Herzke, D., Lohmann, R., Ng, C. A., Trier, X., & Wang, Z. (2020). An overview of the uses of per-and polyfluoroalkyl substances (PFAS). *Environmental Science: Processes & Impacts*, *22*(12), 2345–2373. https://doi.org/10.1039/D0EM00291G

Gobelius, L., Lewis, J., & Ahrens, L. (2017). Plant uptake of per-and polyfluoroalkyl substances at a contaminated fire training facility to evaluate the phytoremediation potential of various plant species. *Environmental Science & Technology*, *51*(21), 12602–12610.

Goldman, P. (1965). The enzymatic cleavage of the carbon-fluorine bond in fluoroacetate. *Journal of Biological Chemistry*, *240*(8), 3434–3438.

Gole, V. L., Fishgold, A., Sierra-Alvarez, R., Deymier, P., & Keswani, M. (2018). Treatment of perfluorooctane sulfonic acid (PFOS) using a large-scale sonochemical reactor. *Separation and Purification Technology*, *194*, 104–110.

Gomez-Ruiz, B., Gómez-Lavín, S., Diban, N., Boiteux, V., Colin, A., Dauchy, X., & Urtiaga, A. (2017). Efficient electrochemical degradation of poly-and perfluoroalkyl substances (PFASs) from the effluents of an industrial wastewater treatment plant. *Chemical Engineering Journal*, *322*, 196–204.

Gomez-Ruiz, B., Ribao, P., Diban, N., Rivero, M.J., Ortiz, I., & Urtiaga, A. (2018). Photocatalytic degradation and mineralization of perfluorooctanoic acid (PFOA) using a composite TiO2 -rGO catalyst. *Journal of Hazardous Materials*, *344*, 950–957. https://10.1016/j.jhazmat.2017.11.048

Gribble, G. W. (2002). Naturally occurring organofluorines. In G. W. Gribble & T. L. Letavic (Eds.), Organofluorines (pp. 121–136). Springer.&

Guo, W., Huo, S., Feng, J., & Lu, X. (2017). Adsorption of perfluorooctane sulfonate (PFOS)on corn straw-derived biochar prepared at different pyrolytic temperatures. *Journal of the Taiwan Institute of Chemical Engineers*, *78*, 265–271. https://doi.org/10.1016/j.jtice.2017.06.013

Hao, F., Guo, W., Wang, A., Leng, Y., & Li, H. (2014). Intensification of sonochemical degradation of ammonium perfluorooctanoate by persulfate oxidant. *Ultrasonics Sonochemistry*, *21*(2), 554–558.

Harding-Marjanovic, K. C., Houtz, E. F., Yi, S., Field, J. A., Sedlak, D. L., & Alvarez-Cohen, L. (2015). Aerobic biotransformation of fluorotelomer thioether amido sulfonate (Lodyne) in AFFF-amended microcosms. *Environmental Science & Technology*, *49*(13), 7666–7674.

Harper, D. B., O'Hagan, D., & Murphy, C. D. (2003). Fluorinated natural products: Occurrence and biosynthesis BT. In G. W. Gribble (Ed.), Natural production of organohalogen compounds (pp. 141–169). Springer Berlin Heidelberg. https://doi.org/10.1007/b10454

Heinicke, G., & Hennig, H.-P. (1984). Tribochemistry. Walter de Gruyter GmbH & Co KG.

Higgins, C. P., & Luthy, R. G. (2006). Sorption of perfluorinated surfactants on sediments. *Environmental Science & Technology*, *40*(23), 7251–7256.

Hori, H., Nagaoka, Y., Murayama, M., & Kutsuna, S. (2008). Efficient decomposition of perfluorocarboxylic acids and alternative fluorochemical surfactants in hot water. *Environmental Science & Technology*, *42*(19), 7438–7443.

Hori, H., Yamamoto, A., Hayakawa, E., Taniyasu, S., Yamashita, N., Kutsuna, S., Kiatagawa, H., & Arakawa, R. (2005). Efficient decomposition of environmentally persistent perfluorocarboxylic acids by use of persulfate as a photochemical oxidant. *Environmental Science & Technology*, *39*(7), 2383–2388.

Hori, H., Yamamoto, A., Koike, K., Kutsuna, S., Osaka, I., & Arakawa, R. (2007). Persulfate-induced photochemical decomposition of a fluorotelomer unsaturated carboxylic acid in water. *Water Research, 41*(13), 2962–2968.

Hu, Z., Song, X., Wei, C., & Liu, J. (2017). Behavior and mechanisms for sorptive removal of perfluorooctane sulfonate by layered double hydroxides. *Chemosphere, 187*, 196–205.

Huang, Q. (2017). *Electrochemcial degradation of perluoroalkyl acids by macroporous titanium suboxide anode.* SERDP ESTCP Symposium Per-And Polyfluoroalkyl Substances (PFAS)-Treatments to Replacements.

Inyang, M., & Dickenson, E. R. V. (2017). The use of carbon adsorbents for the removal of perfluoroalkyl acids from potable reuse systems. *Chemosphere, 184*, 168–175.

Jaffré, T., Kersten, W., Brooks, R. R., & Reeves, R. D. (1979). Nickel uptake by Flacountiaceae of New Caledonia. *Proceedings of the Royal Society of London. Series B. Biological Sciences, 205*(1160), 385–394.

Jin, L., & Zhang, P. (2015). Photochemical decomposition of perfluorooctane sulfonate (PFOS) in an anoxic alkaline solution by 185 nm vacuum ultraviolet. *Chemical Engineering Journal, 280*, 241–247.

Jovicic, V., Khan, M., Zbogar-Rasic, A., Fedorova, N., Poser, A., Swoboda, P., & Delgado, A. (2018). Degradation of low concentrated perfluorinated compounds (PFCs) from water samples using non-thermal atmospheric plasma (NTAP). *Energies, 11*(5). https://10.3390/en11051290

Kerfoot, W.B. (2013). Method and apparatus for treating perfluoroalkyl compounds. Google Patents. Available online: https://patents.google.com/patent/US9694401B2/enA.

Key, B. D., Howell, R. D., & Criddle, C. S. (1998). Defluorination of organofluorine sulfur compounds by Pseudomonas sp. strain D2. *Environmental Science & Technology, 32*(15), 2283–2287.

Kim, T.-H., Lee, S.-H., Kim, H. Y., Doudrick, K., Yu, S., & Kim, S. D. (2019). Decomposition of perfluorooctane sulfonate (PFOS) using a hybrid process with electron beam and chemical oxidants. *Chemical Engineering Journal, 361*, 1363–1370. https://10.1016/j.cej.2018.10.195

Kirsch, P. (2004). Modern Fluoroorganic Chemistry: Synthesis, Reactivity, Applications. WILEY-VCH Verlag GmbH & Co. KGaA, Weinheim.

Kissa, E. (2001). Fluorinated surfactants and repellents. In M. Dekker (Ed.), *Surfactant science series* (2nd ed., Vol. 97), 1–54. CRC Press.

Kucharzyk, K. H., Darlington, R., Benotti, M., Deeb, R., & Hawley, E. (2017). Novel treatment technologies for PFAS compounds: A critical review. *Journal of Environmental Management, 204*(Pt 2), 757–764. https://10.1016/j.jenvman.2017.08.016

Lath, S., Navarro, D. A., McLaughlin, M. J., Kumar, A., & Losic, D. (2017). Adsorption of perfluorooctanoic acid (PFOA) using graphene-based materials. Available at: https://www.battelle.org/docs/default-source/conferences/chlorinated-conference/proceedings/2018-chlorinated-conference-proceedings/a9-pfas-remediation/471.pdf?sfvrsn=75168188_0

Lee, H., Tevlin, A. G., Mabury, S. A., & Mabury, S. A. (2014). Fate of polyfluoroalkyl phosphate diesters and their metabolites in biosolids-applied soil: Biodegradation and plant uptake in greenhouse and field experiments. *Environmental Science & Technology, 48*(1), 340–349.

Lee, Y.-C., Lo, S.-L., Chiueh, P.-T., & Chang, D.-G. (2009). Efficient decomposition of perfluorocarboxylic acids in aqueous solution using microwave-induced persulfate. *Water Research, 43*(11), 2811–2816.

Li Xuchun, Fang, J., Liu, G., Zhang, S., Pan, B., & Ma, J. (2014). Kinetics and efficiency of the hydrated electron-induced dehalogenation by the sulfite/UV process. *Water Research, 62*, 220–228.

Lin, H., Wang, Y., Niu, J., Yue, Z., & Huang, Q. (2015). Efficient sorption and removal of per-fluoroalkyl acids (PFAAs) from aqueous solution by metal hydroxides generated in situ by electrocoagulation. *Environmental Science & Technology, 49*(17), 10562–10569. https://10.1021/acs.est.5b02092

Liu, J., & Avendaño, S. M. (2013). Microbial degradation of polyfluoroalkyl chemicals in the environment: A review. *Environment International, 61*, 98–114.

Lu, D., Sha, S., Luo, J., Huang, Z., & Jackie, X. Z. (2020). Treatment train approaches for the remediation of per- and polyfluoroalkyl substances (PFAS): A critical review. *Journal of Hazardous Materials, 386*, 121963.

Luo, Q., Lu, J., Zhang, H., Wang, Z., Feng, M., Chiang, S.-Y. D., Woodward, D., & Huang, Q. (2015). Laccase-catalyzed degradation of perfluorooctanoic acid. *Environmental Science & Technology Letters, 2*(7), 198–203.

Ma, SH., Wu, MH., Tang, L. et al. (2017). EB degradation of perfluorooctanoic acid and per-fluorooctane sulfonate in aqueous solution. NUCL SCI TECH 28, 137 (2017). https://doi.org/10.1007/s41365-017-0278-8

Mahendra, S., Rome, L. H., Kickhoefer, V. A., & Wang, M. (2016). Bioaugmentation with vaults: Novel in situ remediation strategy for transformation of perfluoroalkyl compounds. SERDP Project ER-2422. Available at: https://apps.dtic.mil/sti/trecms/pdf/AD1022307.pdf

Mahyar, A., Miessner, H., Mueller, S., Hama Aziz, K. H., Kalass, D., Moeller, D., Kretschmer, K., Robles Manuel, S., & Noack, J. (2019). Development and application of different non-thermal plasma reactors for the removal of Perfluorosurfactants in water: A comparative study. *Plasma Chemistry and Plasma Processing, 39*(3), 531–544. https://10.1007/s11090-019-09977-6

Matheson, L. J., Guidetti, J. R., Visscher, P. T., Schaefer, J. K., & Oremland, R. S. (1996). Summary of research results on bacterial degradation of trifluoroacetate (TFA), October, 1993-October, 1995. Open-File Report, 96–219.

McCleaf, P., Englund, S., Östlund, A., Lindegren, K., Wiberg, K., & Ahrens, L. (2017). Removal efficiency of multiple poly-and perfluoroalkyl substances (PFASs) in drinking water using granular activated carbon (GAC) and anion exchange (AE) column tests. *Water Research, 120*, 77–87.

McGuire, M. E., Schaefer, C., Richards, T., Backe, W. J., Field, J. A., Houtz, E., Sedlak, D. L., Guelfo, J. L., Wunsch, A., & Higgins, C. P. (2014). Evidence of remediation-induced alteration of subsurface poly- and perfluoroalkyl substance distribution at a former firefighter training area. *Environmental Science & Technology, 48*(12), 6644–6652. https://doi.org/10.1021/es5006187

MDH. (2019). *Human health-based water guidance table.* Minnesota Department of Health. https://www.health.state.mn.us/communities/environment/risk/guidance/gw/table.html

Merino, N., Qu, Y., Deeb, R. A., Hawley, E. L., Hoffmann, M. R., & Mahendra, S. (2016). Degradation and removal methods for perfluoroalkyl and polyfluoroalkyl substances in water. *Environmental Engineering Science, 33*(9), 615–649.

Milinovic, J., Lacorte, S., Vidal, M., & Rigol, A. (2015). Sorption behaviour of perfluoroal-kyl substances in soils. *Science of the Total Environment, 511*, 63–71. https://10.1016/j.scitotenv.2014.12.017

Mitchell, S. M., Ahmad, M., Teel, A. L., & Watts, R. J. (2014). Degradation of perfluoroocta-noic acid by reactive species generated through catalyzed H2O2 propagation reactions. *Environmental Science & Technology Letters, 1*(1), 117–121.

Natarajan, R., Azerad, R., Badet, B., & Copin, E. (2005). Microbial cleavage of CF bond. *Journal of Fluorine Chemistry, 126*(4), 424–435. https://10.1016/j.jfluchem.2004.12.001

Niu, J., Li, Y., Shang, E., Xu, Z., & Liu, J. (2016). Electrochemical oxidation of perfluorinated compounds in water. *Chemosphere, 146*, 526–538. https://10.1016/j.chemosphere.2015.11.115

Ochoa-Herrera, V., & Sierra-Alvarez, R. (2008). Removal of perfluorinated surfactants by sorption onto granular activated carbon, zeolite and sludge. *Chemosphere, 72*(10), 1588–1593.

Ochoa-Herrera, V., Field, J. A., Luna-Velasco, A., & Sierra-Alvarez, R. (2016). Microbial toxicity and biodegradability of perfluorooctane sulfonate (PFOS) and shorter chain perfluoroalkyl and polyfluoroalkyl substances (PFASs). *Environmental Science: Processes & Impacts, 18*(9), 1236–1246.

OECD. (2008). *OECD glossary of statistical terms.* Organisation for Economic Co-operation and Development.

Oliaei, F., Kriens, D., Weber, R., & Watson, A. (2013). PFOS and PFC releases and associated pollution from a PFC production plant in Minnesota (USA). *Environmental Science and Pollution Research, 20*, 1977–1992.

OPEC. (2017). Fractionation (DFF) solutions. https://opecsystems.com/shop/category/pfas-solutions

Parsons, J. R., Sáez, M., Dolfing, J., & De Voogt, P. (2008). Biodegradation of perfluorinated compounds. *Reviews of Environmental Contamination and Toxicology, 196*, 53–71.

Pennell, K. D., Cápiro, N. L., Fortner, J. D., Arnold, W. A., Simcik, M. F., & Hatton, J. (2017). In situ treatment options for PFAS-contaminated groundwater, 175-177. Proceedings of 7th International Contaminated Site Remediation Conference incorporating the 1st International PFAS Conference: Program and Proceedings, CleanUp 2017 Conference, Melbourne, Australia, 10 - 14 September 2017.

Pillai, S. D. (2016). Introduction to electron-beam food irradiation. *Chemical Engineering Progress, 112*(11), 36–44.

Pillai, S.D., & Shayanfar, S. (2017). Electron Beam Technology and Other Irradiation Technology Applications in the Food Industry. In: Venturi, M., D'Angelantonio, M. (eds) *Applications of Radiation Chemistry in the Fields of Industry, Biotechnology and Environment.* Topics in Current Chemistry Collections. Springer, Cham. https://doi.org/10.1007/978-3-319-54145-7_9

Plumlee, M. H., Larabee, J., & Reinhard, M. (2008). Perfluorochemicals in water reuse. *Chemosphere, 72*(10), 1541–1547.

Plumlee, M. H., McNeill, K., & Reinhard, M. (2009). Indirect photolysis of perfluorochemicals: Hydroxyl radical-initiated oxidation of N-ethyl perfluorooctane sulfonamido acetate (N-EtFOSAA) and other perfluoroalkanesulfonamides. *Environmental Science & Technology, 43*(10), 3662–3668.

Pontius, F. (2019). Regulation of Perfluorooctanoic Acid (PFOA) and Perfluorooctane Sulfonic Acid (PFOS) in Drinking Water: A Comprehensive Review. *Water, 11*(10), 2003. https://doi.org/10.3390/w11102003

Potter, S. (2016). Dynamic groundwater recirculation and adaptive design concepts: Don't accept the status quo. https://www.arcadis.com/media/5/1/E/%7B51EB4FF8-3B01-4948-8F5E-0ACE28C8E861%7DPotter_DGR_TechEx 20

Praveen, C., Jesudhasan, P. R., Reimers, R. S., & Pillai, S. D. (2013). Electron beam inactivation of selected microbial pathogens and indicator organisms in aerobically and anaerobically digested sewage sludge. *Bioresource Technology, 144*, 652–657.

Qiu, Y., Fujii, S., Tanaka, S., & Koizumi, A. (2006). Preliminary study on the treatment of perfluorinated chemicals (PFCs) by advanced oxidation processes (AOPs). *Proceedings of 12th Seminar of JSPS-MOE Core University Program*, Kyoto, Japan, 12, 17–30.

Qu, Y., Zhang, C., Li, F., Chen, J., & Zhou, Q. (2010). Photo-reductive defluorination of perfluorooctanoic acid in water. *Water Research, 44*(9), 2939–2947.

Rattanaoudom, R., Visvanathan, C., & Boontanon, S. K. (2012). Removal of concentrated PFOS and PFOA in synthetic industrial wastewater by powder activated carbon and hydrotalcite. *Journal of Water Sustainability* , 2(4), 245–258.

Redmon, J. H., DeLuca, N. M., Thorp, E., Liyanapatirana, C., Allen, L., Andrew J., & Kondash, A. J. (2025). Hold my beer: The linkage between municipal water and brewing location on PFAS in popular beverages. *Environmental Science & Technology*, 59(17), 8368–8379. https://doi.org/10.1021/acs.est.4c11265

Regenesis. (2017). PlumeStop® Liquid Activated CarbonTM. https://regenesis.com/eur/remediation-products/plumestop-liquid-activated-carbon/

Rodriguez, K. L., Hwang, J.-H., Esfahani, A. R., Sadmani, A. H. M. A., & Lee, W. H. (2020). Recent developments of PFAS-detecting sensors and future direction: A review. *Micromachines*, 11(7), 667.

Rodriguez-Freire, L., Abad-Fernández, N., Sierra-Alvarez, R., Hoppe-Jones, C., Peng, H., Giesy, J. P., Snyder, S., & Keswani, M. (2016). Sonochemical degradation of perfluorinated chemicals in aqueous film-forming foams. *Journal of Hazardous Materials*, 317, 275–283.

Rodriguez-Freire, L., Balachandran, R., Sierra-Alvarez, R., & Keswani, M. (2015). Effect of sound frequency and initial concentration on the sonochemical degradation of perfluorooctane sulfonate (PFOS). *Journal of Hazardous Materials*, 300, 662–669.

Ross, I., Hurst, J., Miles, J., McDonough, J., & Burdick, J. (2017). Remediation of poly- and perfluoro alkyl substances: Developing remediation technologies for emerging challenges. Proceedings of the 7th International Contaminated Site Remediation Conference (CleanUp 2017), Melbourne, VIC (Australia), 10-14 Sep 2017.

Sahu, S. P., Qanbarzadeh, M., Ateia, M., Torkzadeh, H., Maroli, A. S., & Cates, E. L. (2018). Rapid degradation and mineralization of perfluorooctanoic acid by a new petitjeanite Bi3O(OH)(PO4)2 microparticle ultraviolet photocatalyst. *Environmental Science & Technology Letters*, 5(8), 533–538. https://10.1021/acs.estlett.8b00395

Schaefer, C. E., Andaya, C., Urtiaga, A., McKenzie, E. R., & Higgins, C. P. (2015). Electrochemical treatment of perfluorooctanoic acid (PFOA) and perfluorooctane sulfonic acid (PFOS) in groundwater impacted by aqueous film forming foams (AFFFs). *Journal of Hazardous Materials*, 295, 170–175.

Schaefer, C.E., Choyke, S., Ferguson, P. L., Andaya, C., Burant, A., Maizel, A., Strathmann, T.J.,& Higgins, C.P. (2018). Electrochemical Transformations of Perfluoroalkyl Acid (PFAA) Precursors and PFAAs in Groundwater Impacted with Aqueous Film Forming Foams. Environmental Science & Technology, 52(18. https://doi.org/10.1021/acs.est .8b02726.

Schröder, H. F., & Meesters, R. J. W. (2005). Stability of fluorinated surfactants in advanced oxidation processes—A follow up of degradation products using flow injection–mass spectrometry, liquid chromatography–mass spectrometry and liquid chromatography– multiple stage mass spectrometry. *Journal of Chromatography A*, (1), 110–119.

Senevirathna, S., Tanaka, S., Fujii, S., Kunacheva, C., Harada, H., Shivakoti, B. R., & Okamoto, R. (2010). A comparative study of adsorption of perfluorooctane sulfonate (PFOS) onto granular activated carbon, ion-exchange polymers and non-ion-exchange polymers. *Chemosphere*, 80(6), 647–651.

Singh, R. K., Fernando, S., Baygi, S. F., Multari, N., Thagard, S. M., & Holsen, T. M. (2019). Breakdown products from perfluorinated alkyl substances (PFAS) degradation in a plasma-based water treatment process. *Environmental Science & Technology*, 53(5), 2731–2738. https://10.1021/acs.est.8b07031

Song, Z., Tang, H., Wang, N., & Zhu, L. (2013). Reductive defluorination of perfluorooctanoic acid by hydrated electrons in a sulfite-mediated UV photochemical system. *Journal of Hazardous Materials*, 262, 332–338.

Sun, M., Arevalo, E., Strynar, M., Lindstrom, A., Richardson, M., Kearns, B., Pickett, A., Smith, C., & Knappe, D. R. U. (2016). Legacy and emerging perfluoroalkyl substances are important drinking water contaminants in the Cape Fear River Watershed of North Carolina. *Environmental Science & Technology Letters, 3*(12), 415–419.

Suthersan, S., Carroll, P., Schnobrich, M., Horst, J., Potter, S., & Peters, L. (2015a). Cleaning up a 3-mile-long groundwater plume: It can be done. *Groundwater Monitoring and Remediation, 35*(4), 27–35.

Suthersan, S., Killenbeck, E., Potter, S., Divine, C., & LeFrancois, M. (2015b). Resurgence of pump and treat solutions: Directed groundwater recirculation. *Groundwater Monitoring & Remediation, 35*(2), 23–29.

Tang, C. Y., Fu, Q. S., Criddle, C. S., & Leckie, J. O. (2007). Effect of flux (transmembrane pressure) and membrane properties on fouling and rejection of reverse osmosis and nanofiltration membranes treating perfluorooctane sulfonate containing wastewater. *Environmental Science & Technology, 41*(6), 2008–2014. https://doi.org/10.1021/es062052f

Trautmann, A. M., Schell, H., Schmidt, K. R., Mangold, K. M., & Tiehm, A. (2015). Electrochemical degradation of perfluoroalkyl and polyfluoroalkyl substances (PFASs) in groundwater. *Water Science and Technology, 71*(10), 1569–1575. http://10.2166/wst.2015.143

Tseng N, Wang N, Szostek B. & Mahendra S. (2014) Biotransformation of 6:2 fluorotelomer alcohol (6:2 FTOH) by a wood-rotting fungus. Environ Sci Technol. 2014;48(7):4012-20. doi: 10.1021/es4057483.

Urtiaga, A., Fernandez-Castro, P., Gómez, P., & Ortiz, I. (2014). Remediation of wastewaters containing tetrahydrofuran. Study of the electrochemical mineralization on BDD electrodes. *Chemical Engineering Journal, 239*, 341–350.

USEPA. (1997). *High Voltage Environmental Applications, Inc. Electron Beam Technology* (EPA/540/R-96/504). USEPA.

USEPA. (2024a). *Per- and polyfluoroalkyl substances (PFAS) final PFAS national primary drinking water regulation.* USEPA. https://www.epa.gov/sdwa/and-polyfluoroalkyl-substances-pfas

USEPA. (2024b). *PFAS strategic roadmap: EPA's commitments to action 2021–2024.* US.EPA. https://www.epa.gov/pfas/pfas-strategic-roadmap-epas-commitments-action-2021-2024

van Pée, K.-H., & Unversucht, S. (2003). Biological dehalogenation and halogenation reactions. *Chemosphere, 52*(2), 299–312.

Vecitis, Chad D, Park, H., Cheng, J., Mader, B. T., & Hoffmann, M. R. (2009). Treatment technologies for aqueous perfluorooctanesulfonate (PFOS) and perfluorooctanoate (PFOA). *Frontiers of Environmental Science & Engineering in China, 3*, 129–151.

Waite, T. D., Kurucz, C. N., Cooper, W. J., & Brown, D. (1998). Full scale electron beam systems for treatment of water, wastewater and medical waste. Available at: https://inis.iaea.org/records/zkncq-d9t60/files/29050419.pdf?download=1

Wang, L., Batchelor, B., Pillai, S. D., & Botlaguduru, V. S. V. (2016). Electron beam treatment for potable water reuse: Removal of bromate and perfluorooctanoic acid. *Chemical Engineering Journal, 302*, 58–68.

Wang, N., Szostek, B., Buck, R. C., Folsom, P. W., Sulecki, L. M., Capka, V., Berti, W. R., & Gannon, J. T. (2005). Fluorotelomer alcohol biodegradation direct evidence that perfluorinated carbon chains breakdown. *Environmental Science & Technology, 39*(19), 7516–7528.

Watanabe, N., Takemine, S., Yamamoto, K., Haga, Y., & Takata, M. (2016). Residual organic fluorinated compounds from thermal treatment of PFOA, PFHxA and PFOS adsorbed onto granular activated carbon (GAC). *Journal of Material Cycles and Waste Management, 18*, 625–630.

Xiao, L., Ling, Y., Alsbaiee, A., Li, C., Helbling, D. E., & Dichtel, W. R. (2017b). β-Cyclodextrin polymer network sequesters perfluorooctanoic acid at environmentally relevant concentrations. *Journal of the American Chemical Society, 139*(23), 7689–7692.

Xiao, X., Ulrich, B. A., Chen, B., & Higgins, C. P. (2017). Sorption of poly-and perfluoroalkyl substances (PFASs) relevant to aqueous film-forming foam (AFFF)-impacted groundwater by biochars and activated carbon. *Environmental Science & Technology, 51*(11), 6342–6351.

Yu, Q., Deng, S., & Yu, G. (2008). Selective removal of perfluorooctane sulfonate from aqueous solution using chitosan-based molecularly imprinted polymer adsorbents. *Water Research, 42*(12), 3089–3097.

Zaggia, A., Conte, L., Falletti, L., Fant, M., & Chiorboli, A. (2016). Use of strong anion exchange resins for the removal of perfluoroalkylated substances from contaminated drinking water in batch and continuous pilot plants. *Water Research, 91*, 137–146.

Zembouai, I., Kaci, M., Bruzaud, S., Pillin, I., Audic, J.-L., Shayanfar, S., & Pillai, S. D. (2016). Electron beam radiation effects on properties and ecotoxicity of PHBV/PLA blends in presence of organo-modified montmorillonite. *Polymer Degradation and Stability, 132*, 117–126.

Zhang, C., Tang, J., Peng, C., Jin, M. (2016). Degradation of perfluorinated compounds inwastewater treatment plant effluents by electrochemical oxidation with Nano-ZnO coated electrodes. *Journal of Molecular Liquids, 221*, 1145–1150. https://10.1016/j.molliq.2016.06

Zhang, K., Huang, J., Yu, G., Zhang, Q., Deng, S., & Wang, B. (2013). Destruction of perfluorooctane sulfonate (PFOS) and perfluorooctanoic acid (PFOA) by ball milling. *Environmental Science & Technology, 47*(12), 6471–6477.

Zhang, Y., Moores, A., Liu, J., & Ghoshal, S. (2019). New insights into the degradation mechanism of perfluorooctanoic acid by persulfate from density functional theory and experimental data. *Environmental Science & Technology, 53*(15), 8672–8681. https://10.1021/acs.est.9b00797

Zhi, Y., & Liu, J. (2015). Adsorption of perfluoroalkyl acids by carbonaceous adsorbents: Effect of carbon surface chemistry. *Environmental Pollution, 202*, 168–176.

Zhi, Y., & Liu, J. (2016). Surface modification of activated carbon for enhanced adsorption of perfluoroalkyl acids from aqueous solutions. *Chemosphere, 144*, 1224–1232.

Zhi, Y., & Liu, J. (2018). Sorption and desorption of anionic, cationic and zwitterionic polyfluoroalkyl substances by soil organic matter and pyrogenic carbonaceous materials. *Chemical Engineering Journal, 346*, 682–691.

Zhuo, Q., Deng, S., Yang, B., Huang, J., & Yu, G. (2011). Efficient electrochemical oxidation of perfluorooctanoate using a Ti/SnO2-Sb-Bi anode. *Environmental Science & Technology, 45*(7), 2973–2979.

Zhuo, Q., Wang, J., Niu, J., Yang, B., & Yang, Y. (2020). Electrochemical oxidation of perfluorooctane sulfonate (PFOS) substitute by modified boron doped diamond (BDD) anodes. *Chemical Engineering Journal, 379*. https://10.1016/j.cej.2019.122280.

Ziltek. (2017). RemBind®. http://ziltek.com.au/rembind.html

7 Pathways of Human Exposure to Perfluoroalkyl and Polyfluoroalkyl Substances and Health Effects

7.1 INTRODUCTION

The four main ways through that humans are exposed to PFAS include inhaling air, touching contaminated media, contaminated food, and consuming tainted drinking water (Trudel et al., 2008). A quite significant concentration of PFAS pollutants has been noticed in groundwater, soil, marine water, air, drinking water, and freshwater, thereby posing dangers not only to humans but also to aquatic and terrestrial ecosystems (Panieri, Baralic, Djukic-Cosic, Buha Djordjevic, & Saso, 2022).

Because of their "non-stick" and surface-tension-lowering qualities, PFAS frequently change surface chemistry and repel water and oil, reducing stains. The latter covers uses for aqueous film-forming foams (AFFF), semiconductor production auxiliary materials, metal plating, and fluoropolymer production (Fenton et al., 2021). Changes in chemical manufacture can swiftly eliminate direct exposures caused by product usage, but long-term exposures result from PFAS build up in ocean waters and aquatic food chains, as well as AFFF pollution of groundwater (Dassuncao et al., 2017; Zhang, Lee, Niu, & Liu, 2017). Thus, evaluating the causes of temporal variations in serum PFAS concentrations recorded in environmental biomonitoring research requires an understanding of the relative relevance of these various exposure routes and for projecting exposure hazards in the future (Dassuncao et al., 2017; Dassuncao et al., 2018).

The heightened apprehension for marine life, public health, and wildlife ecology derives from the fact that PFAS molecules are thermally and chemically stable, and there are numerous vectors of exposure through which human and biological communities are exposed during their lifetimes. It is imperative to note that these materials are highly mobile, thus spreading widely, and are prone to bioaccumulation and

                          DOI: 10.1201/9781003625537-7

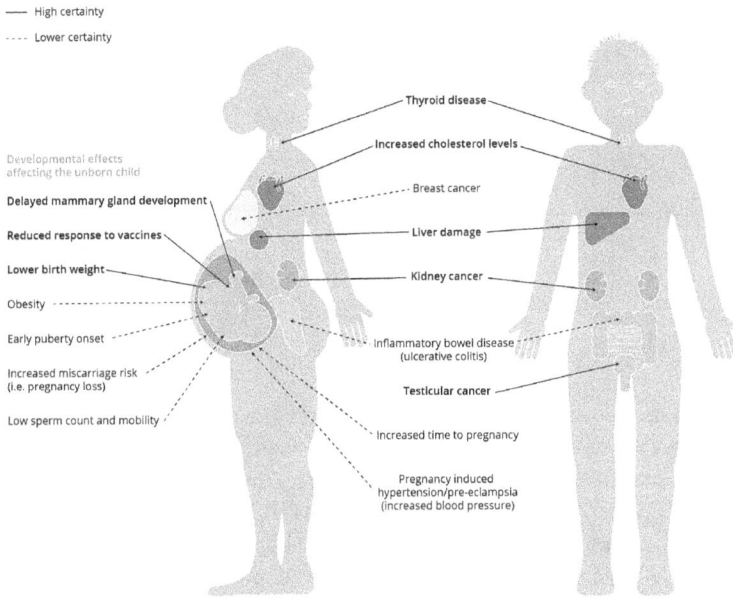

FIGURE 7.1 Effects of PFAS contaminants on humans. Adapted and reproduced from Fenton et al. (2021) with permission from John Wiley & Sons open access article distributed under the terms and conditions of the Creative Commons Attribution (CC BY) license.

biomagnification in biota via contamination of food chains. They have been detected across human biological media (tissue, milk, blood, urine, and other human organs) in various demographic groups within developed countries and are linked to many adverse health effects (Panieri, Baralic, Djukic-Cosic, Buha Djordjevic, & Saso, 2022).

Epidemiological studies assessing the health effects of PFAS have revealed associations with multiple cancers, hypercholesterolemia, compromised immune and liver functions, and considerable birth defects (Bonefeld-Jørgensen, Long, Fredslund, Bossi, & Olsen, 2014) and others as reported and captured in Figure 7.1 (Fenton et al., 2021). Based on this, production companies, decision-makers (both industry and government), and researchers are constantly gaining insights into their overall impact and finding the most ideal mitigation measures to thwart their potential exposure risks.

7.2 EXPOSURE PATHWAYS OF PFAS

Industries, such as the construction of buildings, paints, textiles, firefighting foams, food packaging, and printing inks, mainly produce PFAS pollutants. In addition, further releases result from using and disposing of commercial product categories such as personal care products, textile formulations, household articles, and food-contact substances. Wastewater treatment systems and industrial discharge remain the main contributors to PFAS release. For instance, they allow the discharge of contaminated

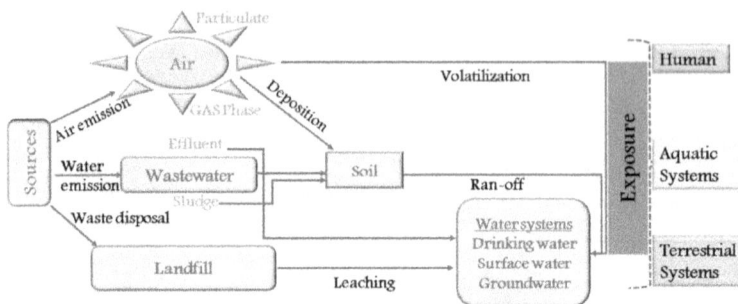

FIGURE 7.2   Illustration of various PFAS pathways.

effluents, polluted sewage sludge, wastewater, and biosolids into the atmosphere and water systems (Fenton et al., 2021). Also, under specific conditions, landfill deposits of PFAS can leach/sink into the soil and enter the groundwater systems, thus exposing humans (Figure 7.2). Given these revelations, it is unsurprising that these substances have become widespread environmental contaminants, adversely affecting public health and the ecosystems through various exposure pathways, which include the atmosphere, vegetables, cereals, milk, fruits, drinking water, and other food supplies (Shahsavari et al., 2021; Wang, Rhodes, Ge, Yu, & Li, 2020; Scher et al., 2018; Awad et al., 2020; Schrenk et al., 2020).

7.3   PFAS OCCURRENCE IN CONSUMER PRODUCTS, INDOOR AIR, AND DUST

The majority of PFAS compounds exhibit minimal evaporation primarily because of their low vapour pressure, low Henry's constant, and elevated boiling point. Nonetheless, other compounds like sulfamidoethanols (FOSEs), alcohols (FTOHs), acrylates (FTACs), and perfluorooctane sulfonates (FOSAs) show volatility and have the potential to partition into the atmospheric phase (Panieri, Baralic, Djukic-Cosic, Buha Djordjevic, & Saso, 2022). Similarly, PFASs are commonly found in textiles (e.g., jackets, upholstery, carpets), construction materials, food packaging materials, impregnation agents, cleaning agents, polishes, paints, and ski waxes (Bečanová, Melymuk, Vojta, Komprdová, & Klánová, 2016; Robel et al., 2017; Schaider et al., 2017; Yuan, Peng, Huang, & Hu, 2016; Zabaleta et al., 2016; Kotthoff, Müller, Jürling, Schlummer, & Fiedler, 2015). PFASs can migrate from food contact papers treated with fluorochemicals into food simulants, including fatty (butter), acidic (vinegar), and aqueous (water and ethanol) media, suggesting a direct pathway for human exposure (Yuan, Peng, Huang, & Hu, 2016). It is believed that there is little dermal exposure to Perfluorooctane sulfonate (PFOS) and Perfluorooctanoic acid (PFOA) from goods (Trudel et al., 2008). In a study by Haug et al (Haug, Huber, Becher, & Thomsen, 2011) that involved 41 Norwegian women, it was shown that although the interior environment (dust, air) may contribute up to around 50% of the PFAS total consumption, food is usually the main exposure channel.

Since numerous consumer items include precursor molecules that the body can biotransform into PFAA, there is further ambiguity about the relevance of exposures from this source (Haug, Huber, Becher, & Thomsen, 2011). Volatile precursors have been detected in indoor situations where utilization and inhalation of products containing PFAS occur (Fromme et al., 2015; Harrad et al., 2010). There has been a rise in the creation of short-chain chemicals and structurally comparable substitute compounds owing to the phase-out of PFOS, PFOA, and their precursors, which requires a more comprehensive method to assess the risk of fluorinated chemicals in humans (Wang, DeWitt, Higgins, & Cousins, 2017). To mitigate this, Robel et al. (2017) estimated the proportion of fluorine that can migrate from certain consumer items and be exposed to humans by measuring overall fluorine concentrations. According to the scientists, up to 16% of the total fluorine assessed by particle-induced gamma-ray emission (PIGE) is not captured by standard PFAS measurement techniques (Robel et al., 2017). Thus, more investigation is required to determine the relationship between PFAS concentrations in goods and those in food, dust, and air, as well as the overall effects of these concentrations on human exposure, mainly among populations with differing product usage patterns.

7.3.1 Occurrence of PFAS in Water Systems

Several PFAS compounds have continuously been detected in water systems and aquifers due to their solubility, especially the ionic PFAS, which has short lengths of carbon chains, and the longer-chain PFAS with 6 to 8 carbon atoms. They are distributed in the water sediment and biota. Since water systems cover two-thirds of the earth, they play a key role in the ecosystems by hosting living organisms, acting as a supply of clean drinking water, and a receptor of hazardous wastes from industries, factories, and homes. Different water systems as sources of PFAS are discussed in the following sections.

7.3.2 Drinking Water

The availability of safe and clean drinking water is universally identified as a basic human right and is captured as a key target within the Sustainable Development Goals (SDGs). Therefore, efforts must be made to reduce and mitigate pollution and restrain the release of pollutants into drinking water, which have been proven to pose serious health risks. Studies indicate that blood concentrations may exceed these levels by up to 100-fold through long-term exposure to PFOA through drinking water (Garnick et al., 2021). Therefore, drinking water is a significant contributor to widespread exposure to PFAS, especially those close to polluted sites (Banzhaf, Filipovic, Lewis, Sparrenbom, & Barthel, 2017; Hu et al., 2016). The US Environmental Protection Agency (EPA) recommended in 2016 that the health-based lifetime water advisory limit for PFOS and PFOA in drinking water be set at 70 ng/L (EPA, 2016). When compared to the reference dose (RfD) that the US EPA used to create the 2016 lifetime advisory, the 2018 Minimum Risk Levels (MRLs) for PFOS and PFOA were further reduced by the US Agency for Toxic Substances and Disease Registry (ATSDR) by nearly an order of magnitude (ATSDR, 2018). According to the ATSDR's MRLs, drinking water warning thresholds for PFOA and PFOS would be 11 ng/L and 7 ng/L, respectively. Some state and international authorities have suggested lifelong drinking water warnings that contain up to 11 or 12 PFAS (ATSDR, 2018). In line with the standard dosage for immunotoxicity linked to PFAS exposure in children in the Faroe Islands, it is projected that the health-based lifetime water advisory limit must not exceed 1 ng/L (Grandjean & Budtz-Jørgensen, 2013).

Sadia et al. conducted a comprehensive study on the occurrence, fate, and related health risks of PFAS in raw and produced drinking water (Sadia et al., 2023). The study found that over 96% of drinking water systems contained detectable levels of PFAS, with median concentrations of PFOA and PFOS exceeding the proposed EPA allowable levels. In the water treatment, reverse osmosis performed better, and activated carbon exhibited reasonable effectiveness. Overall, it was opined that even the treated water has significant exposure concerns, and urgent regulatory action and novel treatment modalities should be utilized.

In the wake of contamination of the Cape Fear River in Wilmington, North Carolina, an investigation for the presence of PFAS was performed by drawing blood samples from 344 individuals aged between six and 86 years. It was observed that over 85% of the samples were found to have 6 fluoroethers with GenX substance found above the reporting limit. The median concentration decreased by 34% to 65% over six months because of controlled wastewater discharge. Also, four legacy PFAS, namely PFHxS, PFOA, PFOS, and PFNA, were present with levels exceeding the US national averages (Kotlarz et al., 2020).

Also, an investigation of PFAS contamination in the River Llobregat in Catalonia, Spain, which is a major drinking water source, was conducted. The study detected 21 different PFAS compounds, with PFHxA and PFBA being the most prevalent and accounting for the majority of the total PFAS burden. The contamination was attributed to both industrial and urban sources, with wastewater plants identified as the main contribution channel (Mussabek et al., 2022).

In another study conducted to investigate PFAS in drinking water and serum of the people of southeast Alaska, researchers found that the drinking water was contaminated with PFAS, with concentrations varying from non-detectable to 120 ng/L. The PFOS and PFHxS were found to be the most copious compounds, confirming a significant source of PFAS near Gustavus (Babayev et al., 2022).

7.3.3 PFAS in Marine Systems and Seafood

Marine ecosystems occupy over 70% of the Earth's surface and support a great deal of ecological biodiversity. However, they also act as sinks for pollutants like PFAS, which are released from various waste sources and dispersed through multiple pathways (Zhang, Lohmann, & Sunderland, 2019).

In the study conducted to investigate the presence and distribution of PFAS in sediment samples from riverine and marine environments in the East China Sea, PFOs, PFHpA, and PFOA had the highest prevalence, with PFOS concentrations being 32.4 ng/g dry weight compared to 34.8 ng/g dry weight for PFAS total concentrations. It was noted that the river inputs are the primary PFAS source since PFAS levels were substantially higher in riverine sediments than in marine sediments. As the contaminants moved further away from the coast, generally, the PFAS concentration decreased due to dilution or degradation processes (Yan, Zhang, Zhou, & Yang, 2015).

A research focus on understanding the spatial and temporal distribution of 24 PFAS compounds in water and sediment samples from Jiulong Estuary-Xiamen Bay in China was conducted. This research revealed that the partition coefficients of PFAS increased in relation to longer carbon chains and higher salinity levels (Wang et al., 2020). In another study, the total concentrations of PFAS in surface sediments of the East China Sea were found to increase over the past decade. Additionally, both diversity and levels of emerging PFAS have been on the rise, alluding to their widespread usage and growing environmental concern in the East China region (Li et al., 2024).

Many groups who eat seafood, such as Inuit males in Greenland who often eat fish and marine animals, have been shown to have elevated blood levels of PFASs (Lindh et al., 2012); whaling men in the Faroe Islands (Weihe et al., serum concentrations of polyfluoroalkyl compounds in Faroese whale meat consumers); and commercial fishery employees in China (Zhou et al., 2014). The levels of PFAS in seafood vary greatly, with the greatest quantities seen close to affected areas (Berger et al., 2009). Unevenness in tissue concentrations between locales and species appears to be mostly caused by environmental concentrations of long-chain chemicals (Stahl et al., 2014). PFSA and long-chain chemicals bioaccumulate more than PFCA and short-chain compounds do (Kelly et al., 2009). However, early research on the bioaccumulation potential of PFAS was limited to tests intended for highly lipophilic compounds, therefore, it did not offer a complete picture of all PFAS now in use (Stahl et al., 2014).

The amount of seafood that contributes to the total exposure to PFAS varies quite a little. Cooking has been demonstrated to lower PFAS concentrations, including PFOS (Del Gobbo et al., 2008). Christensen et al. (Christensen et al., 2017) discovered that between 2007 and 2014, high-frequency fish eaters had higher blood PFAS concentrations in the US National Health and Nutrition Examination Survey. According to a recent assessment by the European Food Safety Authority (EFSA), up to 86% of an adult's dietary exposure to PFAS comes from "fish and other seafood". (Schrenk et al., 2020). Hu et al. (2018) demonstrated that it is possible to determine when seafood is the primary exposure source for PFAS with a chain length of C9 by looking for higher blood concentrations of these compounds in people. This finding was supported by birth cohort data from the Faroe Islands, which demonstrated a high correlation between serum levels of perfluoroundecanoic acid (PFUnDA, C11) and hair mercury concentrations, a potent indicator of seafood diet (Dassuncao et al., 2018). The importance of seafood as an exposure source has risen as a result of shifting production away from legacy PFAS, with concentrations of these chemicals in marine biota keeping pace with this transition (Dassuncao et al., 2018).

7.3.4 OCCURRENCE OF PFAS IN SOIL AND PLANTS

PFAS compounds are usually introduced into the environment via subsurface pathways. They often interact with soil by undergoing processes like adsorption, volatilization, biotransformation, speciation, and uptake before eventually migrating into other environmental compartments such as surface water or groundwater (Panieri, Baralic, Djukic-Cosic, Buha Djordjevic, & Saso, 2022). The main contributors of PFAS in unsaturated soil and plants include atmospheric deposition, the discharge of AFFFs, irrigation with polluted water, and the agricultural usage of biosolids or sludges from municipal systems (Sharifan et al., 2021). More industrialized and advanced countries like the US (13,000 pg/g dw), China (14,000 pg/g dw), and Japan (36,000 pg/g dw) have shown a high degree of soil contamination (Strynar, Lindstrom, Nakayama, Egeghy, & Helfant, 2012).

Atmospheric deposition is a major pathway for PFAS migration into terrestrial and aquatic environments. A study conducted in southwestern Vermont and eastern New York reveals that airborne emissions of PFOA emanating from Teflon coating factories significantly polluted soils and groundwaters across an area of 200 km², which affects over 1,200 residential wells and other water systems in the municipality. Through sampling in the Green Mountain National Forest, PFOA was detected in soils and groundwater with concentrations of up to 100 ppt, and at a distance of even 8 km away from the factories (Schroeder, Bond, & Foley, 2021). Another study that examines the atmospheric transport of PFOA and hexafluoropropylene oxide dimer acid (HFPO-DA) was conducted from a fluoropolymer manufacturing plant in Parkersburg, West Virginia. By focusing on the areas upstream and downwind of the manufacturing plant, 94 surface water and 13 soil samples were collected. All surface water samples were found to have PFOA with concentrations beyond 1,000 ng/L at 13 sites and across an 8-km area surrounding the plant. Similarly, HFPO-DA was recorded with levels of over 100 ng/L up to 6.4 km north of the facility. More importantly, a site which is 28 km north of the facility recorded 143 ng/L and 42 ng/L of PFOA and HFPO, respectively (Galloway et al., 2020).

Studies have also been done to evaluate the presence of PFAS in surface soil via AFFF discharge into the environment. For instance, a study to examine the impact of AFFF from the use of PFAS at the US Air Force has been performed. The soil properties' influence on the movement of PFAS from soil to groundwater has been investigated by utilizing metadata from ongoing site research. The retention of PFAS was found to be affected by total carbon and clay content, with high clay content associated with lower soil-to-groundwater ratios, presumably due to air-water partitioning in more permeable soils. PFAS with fewer carbon such as PFOA, exhibit different behaviour compared to PFOS (Anderson, Adamson, & Stroo, 2019). Also, research to investigate the distribution and behaviour of PFAS at a site contaminated by AFFF was conducted. Groundwater samples were analyzed to check the presence and concentration of different types of PFAS, together with anionic, zwitterionic, and cationic compounds. The anionic compounds of PFAS were the most prevalent with zwitterionic and cationic PFAS also contributing significantly to the contamination. It was revealed that chemical structure and site-specific conditions influence mobility and persistence in the subsurface (Nickerson et al., 2020).

Contaminated irrigation water has also been explored as a possible avenue of affecting agricultural soil and the subsequent reach of PFAS to plants and humans. Environmental and dietary exposure to PFOA and PFOS research was performed in the Nakdong River Delta, South Korea. It was established that the yearly average concentration of PFOA and PFOs was between 0.026 to 0.112 µg/L and 0.818 to 1.364 µg/kg in irrigation water and soil, respectively. Values varying from 0.962 in green onions to less than 0.004 in plums, which have the highest plant uptake, were also reported. Also, the rice and leafy vegetables were found to contribute heavily to the daily uptake of PFOA (0.449 ng/kg) and PFOS (0.140 ng/kg), corresponding to 6.4% and 7.9% for PFOA and PFOS, respectively, in reference to the EFSA dose (Choi et al., 2021).

Another important channel through which soil gets contaminated is through the application of biosolids or fertilizer for agricultural purposes. The effect of the application of biosolids has been investigated by studying agricultural soils that have been subjected to wastewater treatment plant (WWTP) sludge for over a decade in Decatur, Alabama. The soil samples from the sludge showed elevated concentrations of PFDA ($\leq$ 990 ng/g), PFDoA ($\leq$ 530 ng/g), PFOA ($\leq$ 320 ng/g), and PFOS ($\leq$ 410 ng/g). Also, secondary fluorotelomer alcohols (sec-FTOHs) were recorded, which were attributed to the degradation of sec-FTOHs into PFAs (Washington, Yoo, Ellington, Jenkins, & Libelo, 2010). Similarly, soils treated by municipal biosolids were investigated to determine the effect of perfluorochemicals (PFCs) more especially PFOA and PFOs. It was ascertained that PFOS was the most prevalent in both biosolids and amended soils, and their concentrations increased with an increase in biosolid application rates. A possible transformation of the PFC precursors was determined through mass analysis, while a lower leaching potential was confirmed through desorption experiments. The short chain of PFCs indicates that they are more mobile and can migrate vertically through the soil profile, as samples were detected down to 120 cm cm in soil cores (Sepulvado, Blaine, Hundal, & Higgins, 2011).

7.4 HEALTH IMPACTS OF PFAS EXPOSURE

The fact that PFAS has been proven to be hazardous and can bioaccumulate in humans, they are likely to cause far-reaching health problems depending on the exposure circumstances such as time, magnitude, and exposure route. It has been demonstrated that the severity of exposure depends on the health of the person, age, ethnicity, gender, and genetics (Fenton et al., 2021). The major avenues of human exposure routes include food sources, cereals, vegetables, drinking water, milk, and inhalation of indoor air and cereals (Shahsavari et al., 2021; Awad et al., 2020).

The PFAS effect on the biological function of various human organs in both males and females has been well documented. For instance, the endocrine-disruptive impacts have been shown to influence fertility, affect the functioning of the thyroid and mammary gland, and alter body weight. Long-term exposure to PFAS has been linked with a high risk of cancer (prostate, kidney, and testicular) and alteration of cholesterol metabolism or compromised immune system efficacy. Also, developmental effects like accelerated puberty and decreased birth weight in newborns have been reported (Panieri, Baralic, Djukic-Cosic, Buha Djordjevic, & Saso, 2022).

7.4.1 CANCER RISK LINKED TO PFAS EXPOSURE

The PFOA substance has been categorized as potentially carcinogenic by the International Agency for Research on Cancer (IARC), with community drinking PFOA-contaminated water showing greater cancer risk (Sunderland et al., 2019). The exposure of the human and occupational community to PFAS has been associated with various malignancies, such as cancer of the vital organs (liver, testicles, and kidney) since numerous PFAS possess oxidative stress or modulate receptor-mediated effects, which are carcinogenic-like characteristics (Sunderland et al., 2019).

Several studies have examined the carcinogenicity of PFAS, mostly concentrating on PFOA and PFOS. Alongside these two, just the PFHxA has been the subject of animal studies, and no results were found (Klaunig et al., 2015). Chemical workers, communities with tainted drinking water, and the general public are all subjects of PFOS and PFOA human research. For every month spent in the chemical division where the manufacturing of PFOA occurs, there is a reported 3.3-fold increase (95% CI, 1.02–10.6) in the death rate from prostate cancer among occupationally exposed personnel, while the case count is minimal (Gilliland & Mandel, 1993). Subsequent information from this occupational cohort refuted the idea that workplace exposure and cancer incidence or death are related (Raleigh et al., 2014). The most compelling proof of elevated cancer risk has come from research conducted on the general public whose drinking water was PFOA-contaminated. Barry et al. (Barry, Winquist, & Steenland, 2013) revealed a strong correlation between testicular and kidney malignancies and PFOA levels in C8 Health Project participants. The general conclusion from the C8 Health Project is based on these investigations. The general population research provides contradictory results. Eriksen et al. (2009) looked into the relationship between cancer risks linked to PFOA exposure among the general public, and they found no connection between the levels of PFOA or PFOS in plasma and cancers of the prostate, bladder, pancreas, or liver. PFOA has been classified by the IARC as a potential group B carcinogen, which is possibly toxic to humans, while there is no known IARC assessment of PFOS.

7.4.2 NEURODEVELOPMENTAL IMPACTS OF PFAS EXPOSURE

The exposure of PFAS to human foetuses has been shown to have effects on their development, with protracted and undesirable long-term consequences. For instance, a study conducted on five quantities of PFAS systems, namely PPFOS, PFDA, PFUnDA, PFOA, and PENA in human foetuses, placentas, and maternal plasma showed that the concentration of PFOS is elevated in maternal plasma when compared to placentas and other fetal organs. The study demonstrated that these harmful systems tend to migrate from the mother to the foetus, which exhibits a high risk of prenatal exposure, notwithstanding the degree of danger or impact (Mamsen et al., 2017). Similarly, lower average weight and variance in growth at infancy and early childhood stages have been associated with high concentrations of PFOS and PFOA, with the degree of harmfulness varying as they pass from the mother to the foetus at changing efficiency (Gao et al., 2022; Kashino et al., 2020; Spratlen et al., 2019; Winkens, Vestergren, Berger, & Cousins, 2017).

Based on in vitro research, PFOS may contribute to brain endothelial cells' tight junctions to "open", increasing the blood-brain barrier's permeability (Wang et al., 2011). Therefore, there has been considerable interest in learning more about the neurotoxic consequences of exposure to PFAS. Exposure to PFOS, PFOA, and PFHxS at the peak period of fast brain growth in mice has been shown to cause an inability to habituate to a novel environment in laboratory animals (Johansson, Fredriksson, & Eriksson, 2008). Liew et al. (Liew, Goudarzi, & Oulhote, 2018) examined 21 epidemiological studies and established conflicting data about how exposure to PFAS affects neurodevelopment. The health results that were looked at included children's behaviours and attention-deficit/hyperactivity disorder (ADHD), infancy-related developmental achievements, and cognitive functions, which include IQ and other scales. There is a substantial diversity in the techniques and instruments employed to evaluate neurodevelopmental outcomes, and developmental trajectories are highly complex. To prove a connection between PFAS exposure and neurodevelopmental outcomes, further study is required.

7.4.3 Immunotoxicity

The human immune response can be suppressed when PFAS exposure occurs, as these systems affect immune cells by altering cytokine expression. A research study aimed to assess the effect of PFAS exposure on the activation of human T cells, which are an essential part of the immune system, was conducted by Maddalon et al. (2023). The in vitro experimental studies focused on T helper cells, cytotoxic T cells, and natural killer T cells (NKT). It was observed that the PFAS exposure resulted in a reduction in the activation of the three cells (T helper, cytotoxic T, and NKT). This highlights the immunotoxic potential of PFAS since the suppression of T cell activation can compromise the immune system functions, possibly resulting in increased susceptibility to infections and diminished immune surveillance.

A study to evaluate how PFAS affect human immune function, particularly antibody production, was performed using two in vitro models with human peripheral blood mononuclear cells (one T cell-dependent and one T cell-independent) (Iulini et al., 2025). It was found that the long-chain substances, i.e., PFOA, PFOS, PFNA, and PFHxS, substantially suppressed the production of antibodies, with PFOA showing the most pronounced effects, especially in male donors. This non-animal alternative assessment immunotoxicity approach demonstrates that the two in vitro systems can reliably be employed to reflect immune suppression.

7.4.4 ENDOCRINE DISRUPTORS AND KIDNEY DISORDERS

The PFAS substances have been linked to endocrine disruption and kidney dysfunction since they are one of the most common and persistent contaminants available in the environment. As one of the endocrine-disrupting chemicals, PFAS has been found to alter thyroid hormone synthesis, transport, metabolism, and receptor binding, thus affecting thyroid function disorders (Boas, Feldt-Rasmussen, Skakkebæk, & Main, 2006). The effect has been shown to influence the early development of fetal and early childhood stages. Equally, maternal thyroid function alteration significantly affects fetal neurodevelopment, leading to potential cognitive impairments and other related developmental disorders.

Both experimental and computational studies (DFT and molecular dynamics) have been conducted to examine and understand the PFAS binding mechanism with proteins (Rajak & Ganguly, 2023; Peng et al., 2024). An in silico study was conducted recently involving molecular docking studies to explore the possible interactions between PFOS and PFOA and key endogenous antioxidant enzymes (Rajak & Ganguly, 2023). It was established that both PFOS and PFOA exhibited strong affinities towards the catalytic pockets of enzymes, resulting in the formation of strong hydrogen bonds and non-polar hydrophobic associations like van der Waals forces. According to this study, the interactions are presumed to obstruct the functional amino acid residues at the enzymes' active sites, leading to antioxidant function, and thus weakening the enzymatic defence system against reactive oxygen species.

Another study by Peng et al. (2024) used multispectroscopic techniques coupled with molecular dynamics and DFT to investigate the binding affinities of PFAS (HFPO-TA, PFOA, PFO3DA, PFHpA, and DFSA) on Human Serum Albumin (HSA) at 298 K. Based on thermodynamic results, it was reported that PFNA and PFO3DA interaction with HSA are exothermic, which were primarily driven by hydrogen bonding, while PFHpA, DFSA, PFOA, and HFPO-TA were endothermic and mediated by hydrophobic interactions. Analysis of the energy decomposition highlighted that van der Waals and electrostatic interactions dominate PFAS binding with DFT, effectively characterizing the PFAS-HAS interactions. Therefore, this study provides good insights into how PFAS interacts with blood proteins and offers a platform to study strategies to mitigate the health effects of these contaminants.

7.5 CONCLUSION

PFAS has appeared as a substantial danger to the natural environment and public health due to their omnipresence in ecological systems and widespread usage, hence, causing adverse health effects. With their pathways to humans being through drinking polluted water, consuming contaminated food, contact with household products, and inhaling air, these exposures to humans underscore their ability to accumulate in biological systems.

Studies on the health effects of PFAS highlight a troubling array of outcomes, which include neurodevelopmental defects, metabolic disorders, compromised immunity, and chronic diseases like cancer. With bioaccumulation, vulnerable populations of infants, pregnant mothers, and exposed occupational workers risk a potential interference with development and physiological processes.

In conclusion, awareness of the pathways of PFAS exposure and their related health risks is indispensable in safeguarding human and environmental health. Therefore, this calls for a collaborative effort among policymakers and industries to formulate policies and any other necessary mitigation measures to address emerging challenges from PFAS, thus ensuring a safer future for future generations.

REFERENCES

Anderson, R., Adamson, D., & Stroo, H. (2019). Partitioning of poly- and perfluoroalkyl substances from soil to groundwater within aqueous film-forming foam source zones. *Journal of contaminant hydrology, 220,* 59–65.

ATSDR. (2018). *Toxicological profile for perfluoroalkyls: Draft for public comment.* Division of Toxicology and Human Health Sciences.

Awad, R., Zhou, Y., Nyberg, E., Namazkar, S., Yongning, W., Xiao, Q., ... Benskin, J. (2020). Emerging per- and polyfluoroalkyl substances (PFAS) in human milk from Sweden and China. Environmental Science. *Environmental Science: Processes & Impacts, 22*(10), 2023–2030.

Babayev, M., Capozzi, S., Miller, P., McLaughlin, K., Medina, S., Byrne, S., ... Salamova, A. (2022). PFAS in drinking water and serum of the people of a southeast Alaska community: A pilot study. *Environmental Pollution, 305*(1), 19246.

Banzhaf, S., Filipovic, M., Lewis, J., Sparrenbom, C., & Barthel, R. (2017). A review of contamination of surface-, ground-, and drinking water in Sweden by perfluoroalkyl and polyfluoroalkyl substances (PFASs). *Ambio, 46*(3), 335–346.

Barry, V., Winquist, A., & Steenland, K. (2013). Perfluorooctanoic acid (PFOA) exposures and incident cancers among adults living near a chemical plant. *Environmental health perspectives, 121*(11–12), 1313–1318.

Bečanová, J., Melymuk, L., Vojta, Š., Komprdová, K., & Klánová, J. (2016). Screening for perfluoroalkyl acids in consumer products, building materials and wastes. *Chemosphere, 164,* 322–329.

Berger, U., Glynn, A., Holmström, K., Berglund, M., Ankarberg, E., & Törnkvist, A. (2009). Fish consumption as a source of human exposure to perfluorinated alkyl substances in Sweden–Analysis of edible fish from Lake Vättern and the Baltic Sea. *Chemosphere, 76*(6), 799–804.

Boas, M., Feldt-Rasmussen, U., Skakkebæk, N., & Main, K. (2006). Environmental chemicals and thyroid function. *European Journal of Endocrinology, 154*(5), 599–611.

Bonefeld-Jørgensen, E., Long, M., Fredslund, S., Bossi, R., & Olsen, J. (2014). Breast cancer risk after exposure to perfluorinated compounds in Danish women: A case–control study nested in the Danish National Birth Cohort. *Cancer Causes & Control, 25,* 1439–1448.

Choi, G., Lee, D., Bruce-Vanderpuije, P., Song, A., Lee, H., Park, S., ... Kim, J. (2021). Environmental and dietary exposure of perfluorooctanoic acid and perfluorooctanesulfonic acid in the Nakdong River, Korea. *Environmental Geochemistry and Health, 43,* 347–360.

Christensen, K., Raymond, M., Blackowicz, M., Liu, Y., Thompson, B., Anderson, H., & Turyk, M. (2017). Perfluoroalkyl substances and fish consumption. *Environmental Research, 154,* 145–151.

Dassuncao, C., Hu, X., Nielsen, F., Weihe, P., Grandjean, P., & Sunderland, E. (2018). Shifting global exposures to poly- and perfluoroalkyl substances (PFASs) evident in longitudinal birth cohorts from a seafood-consuming population. *Environmental Science & Technology, 52*(6), 3738–3747.

Dassuncao, C., Hu, X., Zhang, X., Bossi, R., Dam, M., Mikkelsen, B., & Sunderland, E. (2017). Temporal shifts in poly- and perfluoroalkyl substances (PFASs) in North Atlantic pilot whales indicate large contribution of atmospheric precursors. *Environmental Science & Technology, 51*(8), 4512–4521.

Del Gobbo, L., Tittlemier, S., Diamond, M., Pepper, K., Tague, B., Yeudall, F., & Vanderlinden, L. (2008). Cooking decreases observed perfluorinated compound concentrations in fish. *Journal of Agricultural and Food Chemistry, 56*(16), 7551–7559.

EPA. (2016). Lifetime health advisories and health effects support documents for perfluorooctanoic acid and perfluorooctane sulfonate. *Federal Register, 81*(101), 33250–33251.

Eriksen, K., Sørensen, M., McLaughlin, J., Lipworth, L., Tjønneland, A., Overvad, K., & Raaschou-Nielsen, O. (2009). Perfluorooctanoate and perfluorooctanesulfonate plasma levels and risk of cancer in the general Danish population. *Journal of the National Cancer Institute, 101*(8), 605–609.

Fenton, S., Ducatman, A., Boobis, A., DeWitt, J., Lau, C. N., Smith, J., & Roberts, S. (2021). Per- and polyfluoroalkyl substance toxicity and human health review: Current state of knowledge and strategies for informing future research. *Environmental Toxicology and Chemistry, 40*(3), 606–630.

Fromme, H., Dreyer, A., Dietrich, S., Fembacher, L., Lahrz, T., & Völkel, W. (2015). Neutral polyfluorinated compounds in indoor air in Germany–The LUPE 4 study. *Chemosphere, 139*, 572–578.

Galloway, J., Moreno, A., Lindstrom, A., Strynar, M., Newton, S., May, A., & Weavers, L. (2020). Evidence of air dispersion: HFPO–DA and PFOA in Ohio and West Virginia surface water and soil near a fluoropolymer production facility. *Environmental Science & Technology, 54*(12), 7175–7184.

Gao, Y., Luo, J., Zhang, Y., Pan, C., Ren, Y., Zhang, J.,… Cohort, S. B. (2022). Prenatal exposure to per- and polyfluoroalkyl substances and child growth trajectories in the first two years. *Environmental Health Perspectives, 130*(3), 037006.

Garnick, L., Massarsky, A., Mushnick, A., Hamaji, C., Scott, P., & Monnot, A. (2021). An evaluation of health-based federal and state PFOA drinking water guidelines in the United States. *Science of the Total Environment, 761*, 144107.

Gilliland, F., & Mandel, J. (1993). Mortality among employees of a perfluorooctanoic acid production plant. *Journal of Occupational Medicine, 35*, 950–954.

Grandjean, P., & Budtz-Jørgensen, E. (2013). Immunotoxicity of perfluorinated alkylates calculation of benchmark doses based on serum concentrations in children. *Environmental Health, 12*, 1–7.

Harrad, S., de Wit, C., Abdallah, M., Bergh, C., Bjorklund, J., Covaci, A., … Leonards, P. (2010). Indoor contamination with hexabromocyclododecanes, polybrominated diphenyl ethers, and perfluoroalkyl compounds: An important exposure pathway for people? *Environmental Science & Technology, 44*(9), 3221–3231.

Haug, L., Huber, S., Becher, G., & Thomsen, C. (2011). Characterisation of human exposure pathways to perfluorinated compounds—comparing exposure estimates with biomarkers of exposure. *Environment International, 37*(4), 687–693.

Hu, X., Andrews, D., Lindstrom, A., Bruton, T., Schaider, L., Grandjean, P., … Higgins, C. (2016). Detection of poly- and perfluoroalkyl substances (PFASs) in US drinking water linked to industrial sites, military fire training areas, and wastewater treatment plants. *Environmental Science & Technology Letters, 3*(10), 344–350.

Hu, X., Dassuncao, C., Zhang, X., Grandjean, P., Weihe, P., Webster, G., ... Sunderland, E. (2018). Can profiles of poly- and perfluoroalkyl substances (PFASs) in human serum provide information on major exposure sources? *Environmental Health, 17*, 1–15.

Iulini, M., Bettinsoli, V., Maddalon, A., Galbiati, V., Janssen, A., Beekmann, K., ... Corsini, E. (2025). In vitro approaches to investigate the effect of chemicals on antibody production: The case study of PFASs. *Archives of Toxicology, 99*, 1–12.

Johansson, N., Fredriksson, A., & Eriksson, P. (2008). Neonatal exposure to perfluorooctane sulfonate (PFOS) and perfluorooctanoic acid (PFOA) causes neurobehavioural defects in adult mice. *Neurotoxicology, 29*(1), 160–169.

Kashino, I., Sasaki, S., Okada, E., Matsuura, H., Goudarzi, H., Miyashita, C., ... Kishi, R. (2020). Prenatal exposure to 11 perfluoroalkyl substances and fetal growth: A large-scale, prospective birth cohort study. *Environment International, 136*, 105355.

Kelly, B., Ikonomou, M., Blair, J., Surridge, B., Hoover, D., Grace, R., & Gobas, F. (2009). Perfluoroalkyl contaminants in an Arctic marine food web: Trophic magnification and wildlife exposure. *Environmental Science & Technology, 43*(11), 4037–4043.

Klaunig, J., Shinohara, M., Iwai, H., Chengelis, C., Kirkpatrick, J., Wang, Z., & Bruner, R. (2015). Evaluation of the chronic toxicity and carcinogenicity of perfluorohexanoic acid (PFHxA) in Sprague-Dawley rats. *Toxicologic pathology, 43*(2), 209–220.

Kotlarz, N., McCord, J., Collier, D., Lea, C., Strynar, M., Lindstrom, A., ... Polera, M. (2020). Measurement of novel, drinking water-associated PFAS in blood from adults and children in Wilmington, North Carolina. Environmental health perspectives. *Environmental Health Perspectives, 128*(7), 077005.

Kotthoff, M., Müller, J., Jürling, H., Schlummer, M., & Fiedler, D. (2015). Perfluoroalkyl and polyfluoroalkyl substances in consumer products. *Environmental Science and Pollution Research, 22*, 14546–14559.

Li, Q., Liu, C., Wang, S., Liu, Y., Ma, X., Li, Y., ... Wang, X. (2024). Decade-long historical shifts in legacy and emerging per- and polyfluoroalkyl substances (PFAS) in surface sediments of China's marginal seas: Ongoing production and ecological risks. *Environmental Research, 263*, 119978.

Liew, Z., Goudarzi, H., & Oulhote, Y. (2018). Developmental exposures to perfluoroalkyl substances (PFASs): An update of associated health outcomes. *Current Environmental Health Reports, 5*(1), 1–19. PMID: 29556975. CrossRef View in Scopus, 1–19.

Lindh, C., Rylander, L., Toft, G., Axmon, A., Rignell-Hydbom, A., Giwercman, A., ... Vermeulen, R. (2012). Blood serum concentrations of perfluorinated compounds in men from Greenlandic Inuit and European populations. *Chemosphere, 88*(11), 1269–1275.

Maddalon, A., Pierzchalski, A., Kretschmer, T., Bauer, M., Zenclussen, A., Marinovich, M., ... Herberth, G. (2023). Mixtures of per- and poly-fluoroalkyl substances (PFAS) reduce the in vitro activation of human T cells and basophils. *Chemosphere, 336*, 139204.

Mamsen, L., Jönsson, B., Lindh, C., Olesen, R., Larsen, A., Ernst, E., ... Andersen, C. (2017). Concentration of perfluorinated compounds and cotinine in human foetal organs, placenta, and maternal plasma. *Science of the Total Environment, 596*, 97–105.

Mussabek, D., Söderman, A., Imura, T., Persson, K., Nakagawa, K., Ahrens, L., & Berndtsson, R. (2022). PFAS in the drinking water source: Analysis of the contamination levels, origin and emission rates. *Water, 15*(1), 137.

Nickerson, A., Rodowa, A., Adamson, D., Field, J., Kulkarni, P., Kornuc, J., & Higgins, C. (2020). Spatial trends of anionic, zwitterionic, and cationic PFASs at an AFFF-impacted site. *Environmental Science & Technology, 55*(1), 313–323.

Panieri, E., Baralic, K., Djukic-Cosic, D., Buha Djordjevic, A., & Saso, L. (2022). PFAS molecules: A major concern for the human health and the environment. *Toxics, 10*(2), 44.

Peng, M., Xu, Y., Wu, Y., Cai, X., Zhang, W., Zheng, L., ... Fu, J. (2024). Binding affinity and mechanism of six PFAS with human serum albumin: Insights from multi-spectroscopy, DFT and molecular dynamics approaches. *Toxics, 12*(1), 43.

Rajak, P., & Ganguly, A. (2023). The ligand-docking approach explores the binding affinity of PFOS and PFOA for major endogenous antioxidants: A potential mechanism to fuel oxidative stress. *Sustainable Chemistry for the Environment, 4*, 100047.

Raleigh, K., Alexander, B., Olsen, G., Ramachandran, G., Morey, S., Church, T., ... Allen, E. (2014). Mortality and cancer incidence in ammonium perfluorooctanoate production workers. *Occupational and Environmental Medicine, 71*(7), 500–506.

Robel, A., Marshall, K., Dickinson, M., Lunderberg, D., Butt, C., Peaslee, G., ... Field, J. (2017). Closing the mass balance on fluorine on papers and textiles. *Environmental Science & Technology, 51*(16), 9022–9032.

Sadia, M., Nollen, I., Helmus, R., Ter Laak, T., Béen, F., Praetorius, A., & van Wezel, A. (2023). Occurrence, fate, and related health risks of PFAS in raw and produced drinking water. *Environmental Science & Technology, 57*(8), 3062–3074.

Schaider, L., Balan, S., Blum, A., Andrews, D., Strynar, M., Dickinson, M., ... Peaslee, G. (2017). Fluorinated compounds in US fast food packaging. *Environmental Science & Technology Letters, 4*(3), 105–111.

Scher, D., Kelly, J., Huset, C., Barry, K., Hoffbeck, R., Yingling, V., & Messing, R. (2018). Occurrence of perfluoroalkyl substances (PFAS) in garden produce at homes with a history of PFAS-contaminated drinking water. *Chemosphere, 196*, 548–555.

Schrenk, D., Bignami, M., Bodin, L., Chipman, J., del Mazo, J., Grasl-Kraupp, B., ... Nebbia, C. (2020). Risk to human health related to the presence of perfluoroalkyl substances in food. *EFSA Journal, 18*(9), 06223.

Schroeder, T., Bond, D., & Foley, J. (2021). PFAS soil and groundwater contamination via industrial airborne emission and land deposition in SW Vermont and Eastern New York State, USA. *Environmental Science: Processes & Impacts, 23*(2), 291–301.

Sepulvado, J., Blaine, A., Hundal, L., & Higgins, C. (2011). Occurrence and fate of perfluorochemicals in soil following the land application of municipal biosolids. *Environmental Science & Technology, 45*(19), 8106–8112.

Shahsavari, E., Rouch, D., Khudur, L., Thomas, D., Aburto-Medina, A., & Ball, A. (2021). Challenges and current status of the biological treatment of PFAS-contaminated soils. *Frontiers in Bioengineering and Biotechnology, 8*, 602040.

Sharifan, H., Bagheri, M., Wang, D., Burken, J., Higgins, C., Liang, Y., ... Blotevogel, J. (2021). Fate and transport of per- and polyfluoroalkyl substances (PFASs) in the vadose zone. *Science of the Total Environment, 771*(14), 145427.

Spratlen, M., Perera, F., Lederman, S., Robinson, M., Kannan, K., Trasande, L., & Herbstman, J. (2019). Cord blood perfluoroalkyl substances in mothers exposed to the World Trade Center disaster during pregnancy. *Environmental Pollution, 246*, 482–490.

Stahl, L., Snyder, B., Olsen, A., Kincaid, T., Wathen, J., & McCarty, H. (2014). Perfluorinated compounds in fish from US urban rivers and the Great Lakes. *Science of the Total Environment, 499*, 185–195.

Strynar, M., Lindstrom, A., Nakayama, S., Egeghy, P., & Helfant, L. (2012). Pilot scale application of a method for the analysis of perfluorinated compounds in surface soils. *Chemosphere, 86*(3), 252–257.

Sunderland, E., Hu, X., Dassuncao, C., Tokranov, A., Wagner, C., & Allen, J. (2019). A review of the pathways of human exposure to poly-and perfluoroalkyl substances (PFASs) and present understanding of health effects. *Journal of Exposure Science & Environmental Epidemiology, 29*(2), 131–147.

Trudel, D., Horowitz, L., Wormuth, M., Scheringer, M., Cousins, I., & Hungerbühler, K. (2008). Estimating consumer exposure to PFOS and PFOA. Risk Analysis. *An International Journal, 28*(2), 251–269.

Wang, S., Ma, L., Chen, C., Li, Y., Wu, Y., Liu, Y., … Wang, X. (2020). Occurrence and partitioning behaviour of per- and polyfluoroalkyl substances (PFASs) in water and sediment from the Jiulong Estuary-Xiamen Bay, China. *Chemosphere, 238*, 124578.

Wang, W., Rhodes, G., Ge, J., Yu, X., & Li, H. (2020). Uptake and accumulation of per- and polyfluoroalkyl substances in plants. *Chemosphere, 261*, 127584.

Wang, X., Li, B., Zhao, W., Liu, Y., Shang, D., Fang, W., & Chen, Y. (2011). Perfluorooctane sulfonate triggers tight junction "opening" in brain endothelial cells via phosphatidylinositol 3-kinase. *Biochemical and Biophysical Research Communications, 410*(2), 258–263.

Wang, Z., DeWitt, J., Higgins, C., & Cousins, I. (2017). A never-ending story of per- and polyfluoroalkyl substances (PFASs). *Environmental Science & Technology Journal - ACS Publications, 51*, 2508–2518.

Washington, J., Yoo, H., Ellington, J., Jenkins, T., & Libelo, E. (2010). Concentrations, distribution, and persistence of perfluoroalkylates in sludge-applied soils near Decatur, Alabama, USA. *Environmental Science & Technology, 44*(22), 8390–8396.

Weihe, P., Kato, K., Calafat, A., Nielsen, F., Wanigatunga, A., Needham, L., & Grandjean, P. (2008). Serum concentrations of polyfluoroalkyl compounds in Faroese whale meat consumers. *Environmental Science & Technology, 42*(16), 6291–6295.

Winkens, K., Vestergren, R., Berger, U., & Cousins, I. (2017). Early life exposure to per- and polyfluoroalkyl substances (PFASs): A critical review. *Emerging Contaminants, 3*(2), 55–68.

Yan, H., Zhang, C., Zhou, Q., & Yang, S. (2015). Occurrence of perfluorinated alkyl substances in sediment from estuarine and coastal areas of the East China Sea. *Environmental Science and Pollution Research, 22*, 1662–1669.

Yuan, G., Peng, H., Huang, C., & Hu, J. (2016). Ubiquitous occurrence of fluorotelomer alcohols in eco-friendly paper-made food-contact materials and their implication for human exposure. *Environmental Science & Technology, 50*(2), 942–950.

Zabaleta, I., Bizkarguenaga, E., Bilbao, D., Etxebarria, N., Prieto, A., & Zuloaga, O. (2016). Fast and simple determination of perfluorinated compounds and their potential precursors in different packaging materials. *Talanta, 152*, 353–363.

Zhang, L., Lee, L., Niu, J., & Liu, J. (2017). Kinetic analysis of aerobic biotransformation pathways of a perfluorooctane sulfonate (PFOS) precursor in distinctly different soils. *Environmental Pollution, 229*, 159–167.

Zhang, X., Lohmann, R., & Sunderland, E. (2019). Poly- and perfluoroalkyl substances in seawater and plankton from the northwestern Atlantic margin. *Environmental Science & Technology, 53*(21), 12348–12356.

Zhou, Z., Shi, Y., Vestergren, R., Wang, T., Liang, Y., & Cai, Y. (2014). Highly elevated serum concentrations of perfluoroalkyl substances in fishery employees from Tangxun Lake, China. *Environmental Science & Technology, 48*(7), 3864–3874.

8 Knowledge Gaps and Future Research Directions for Perfluoroalkyl and Polyfluoroalkyl Substances

8.1 INTRODUCTION TO THE KNOWLEDGE GAPS AND FUTURE RESEARCH DIRECTIONS FOR PFASS

The absence of toxicity data and the difficulties in determining the overall number of individual PFASs detected in environmental samples underline the need for knowledge and resources to better comprehend novel and emerging fluorinated substances. EOF (unquantified persistent flame-causing compounds) assess the existence of all flammable organofluorine chemicals (Abunada, Alazaiza, & Bashir, 2020). Research studies conducted in two German towns between 1982 and 2009 have determined that in human plasma samples, between 51% and 100% of the EOF was attributable to quantifiable PFASs (Göckener, Weber, Rüdel, Bücking, & Kolossa-Gehring, 2020). In one city, after 2000, the percentage and quantity of unknown organofluorine in human plasma rose. The hypothesis of this study is in line with the literature on environmental exposure and suggests that people are exposed to several novel and undiscovered organofluorine chemicals (Aro, Eriksson, Karrman, & Yeung, 2021; Rotander et al., 2015; EFSA Panel on Contaminants in the Food Chain [CONTAM] et al., 2018).

How harmful new and developing PFASs are to ecosystems and people is not well known. This is important since it is commonly believed that to participate in risk mitigation efforts, it is required to quantify the possible health risks associated with exposures in places with high PFAS concentrations. Manufacturers of chemicals have asserted that shorter-chain homologues, which are characterized by reduced biological half-lives in humans, have demonstrated low bioaccumulation potential and that replacement PFASs are not linked to any negative health impacts (Bălan, Mathrani, Guo, & Algazi, 2021; Zheng, Eick, & Salamova, 2023). Ongoing research,

DOI: 10.1201/9781003625537-8

however, indicates that shorter-chain compounds may be more able to interact with biomolecules than their longer-chain homologues because they face less steric hindrance (Solan & Lavado, 2023; Hussain et al., 2025). For example, adding oxygen molecules is expected to shorten the carbon chains in perfluoroalkyl ether carboxylic acids (PFECAs), an essential new group of PFASs (Fang, Meng, Schaefer, & Knappe, 2023). The ammonium salt of perfluoro-2-propoxypropanoic acid (PFECA), commercialized under the brand name "GenX" since 2010, is a popular alternative to PFOA. According to a recent hazard evaluation, GenX is more dangerous than PFOA regarding internal dose after factoring in toxicokinetic variability (Satbhai, Vogs, & Crago, 2022). Studies have challenged the application of perfluorinated compounds, instead advocating for regulation based on chemical class properties to regulate them because of the significant environmental persistence, bioaccumulation, and likely toxicity of the whole PFAS family (Bilela et al., 2023).

In conclusion, additional research is needed to understand the periods of exposure to legacy PFASs connected to drinking water and seafood contamination and the exposure pathways and health impacts associated with novel PFAS. Risk mitigation solutions rely on novel technology to reduce PFAS concentrations at polluted locations and drinking water sources. Belated response to previous PFASs has caused extensive human exposure to dangers and effects; lessons should be drawn from this history, but not replicated for newfound PFASs now on the market (Fenton et al., 2021). Further research is required to understand the consequences of PFAS exposures on human health, mostly at critical life stages, and this should not be used as an excuse to postpone risk reduction efforts. PFOS and its precursors underwent a phase-out from 2000 to 2002, which proved the potential benefits of coordinated global action, and it was highly effective in rapidly decreasing the levels of these chemicals to which humans and wildlife are exposed.

This structure is a basis for a full review paper on Perfluoroalkyl and Polyfluoroalkyl Substances. It addresses various topics, including sources and production, environmental destiny and transit, health consequences, legislation and recommendations, analytical methods and monitoring, remediation options, risk communication, and future research objectives. The review aims to synthesize current information on PFAS and identify key areas for additional inquiry and appropriate management of these persistent pollutants.

8.2   CONCLUSION ON THE KNOWLEDGE GAPS AND FUTURE RESEARCH DIRECTIONS FOR PFASS

The growing abundance of new and legacy PFAS churned into the environment daily highlights the significant knowledge gaps in areas such as understanding their toxicity, their sources, human exposure pathways, and short-term and long-term health risks. With a shift to using short-chain and alternative PFAS, studies have shown they are equally unsafe. Therefore, there is an urgent need to provide class-based regulations, advocate for advanced analytical methods, and implement effective mitigation and remediation strategies. Future research should focus on offering a comprehensive risk assessment and synergistic global cooperation towards addressing these harmful contaminants.

REFERENCES

Abunada, Z., Alazaiza, M., & Bashir, M. (2020). An Overview of per- and polyfluoroalkyl substances (PFAS) in the environment: Source, fate, risk and regulations. *Water, 12*(12), 3590.

Aro, R., Eriksson, U., Karrman, A., & Yeung, L. (2021). Organofluorine mass balance analysis of whole blood samples in relation to gender and age. *Environmental Science & Technology, 55*(19), 13142–13151.

Bălan, S., Mathrani, V., Guo, D., & Algazi, A. (2021). Regulating PFAS as a chemical class under the California Safer Consumer Products Program. *Environmental Health Perspectives, 129*(2), 025001.

Bilela, L., Matijošytė, I., Krutkevičius, J., Alexandrino, D., Safarik, I., Burlakovs, J., ... Carvalho, M. (2023). Impact of per- and polyfluorinated alkyl substances (PFAS) on the marine environment: Raising awareness, challenges, legislation, and mitigation approaches under the One Health concept. *Marine Pollution Bulletin, 194*, 115309.

EFSA Panel on Contaminants in the Food Chain (CONTAM), K. H., Alexander, J., Barregård, L., Bignami, M., Brüschweiler, B., Ceccatelli, S., ... Grasl-Kraupp, B. (2018). Risk to human health related to the presence of perfluorooctane sulfonic acid and perfluorooctanoic acid in food. *EFSA Journal, 16*(12), 05194.

Fang, Y., Meng, P., Schaefer, C., & Knappe, D. (2023). Removal and destruction of perfluoroalkyl ether carboxylic acids (PFECAs) in an anion exchange resin and electrochemical oxidation treatment train. *Water Research, 230*, 119522.

Fenton, S., Ducatman, A., Boobis, A., DeWitt, J. L., Ng, C., Smith, J., & Roberts, S. (2021). Per- and polyfluoroalkyl substance toxicity and human health review: Current state of knowledge and strategies for informing future research. *Environmental Toxicology and Chemistry, 40*(3), 606–630.

Göckener, B., Weber, T., Rüdel, H., Bücking, M., & Kolossa-Gehring, M. (2020). Human biomonitoring of per- and polyfluoroalkyl substances in German blood plasma samples from 1982 to 2019. *Environment International, 145*, 106123.

Hussain, H., Jilani, M., Imtiaz, F., Ahmed, T., Arshad, M., Mudassar, M., & Sharif, M. (2025). Advances in the removal of polyfluoroalkyl substances (PFAS) from water using destructive and non-destructive methods. *Green Analytical Chemistry*, 12, 100225.

Rotander, A., Karrman, A., Toms, L., Kay, M. &Mueller, J.(2015). Novel fluorinated surfactants tentatively identified in firefighters using liquid chromatography quadrupole time-of-flight tandem mass spectrometry and a case-control approach. *Environmental Science & Technology, 49*(4), 2434–2442.

Satbhai, K., Vogs, C., & Crago, J. (2022). Comparative toxicokinetics and toxicity of PFOA and its replacement GenX in the early stages of zebrafish. *Chemosphere, 308*, 136131.

Solan, M., & Lavado, R. (2023). Effects of short-chain per- and polyfluoroalkyl substances (PFAS) on human cytochrome P450 (CYP450) enzymes and human hepatocytes: An in vitro study. *Current Research in Toxicology, 5*, 100116.

Zheng, G., Eick, S., & Salamova, A. (2023). Elevated levels of ultrashort-and short-chain perfluoroalkyl acids in US homes and people. *Environmental Science & Technology, 57*(42), 15782–15793.

Index

For Product Safety Concerns and Information please contact our EU
representative GPSR@taylorandfrancis.com
Taylor & Francis Verlag GmbH, Kaufingerstraße 24, 80331 München, Germany